SpringerBriefs in Earth Sciences

SpringerBriefs in Earth Sciences present concise summaries of cutting-edge research and practical applications in all research areas across earth sciences. It publishes peer-reviewed monographs under the editorial supervision of an international advisory board with the aim to publish 8 to 12 weeks after acceptance. Featuring compact volumes of 50 to 125 pages (approx. 20,000–70,000 words), the series covers a range of content from professional to academic such as:

- timely reports of state-of-the art analytical techniques
- bridges between new research results
- snapshots of hot and/or emerging topics
- literature reviews
- in-depth case studies

Briefs will be published as part of Springer's eBook collection, with millions of users worldwide. In addition, Briefs will be available for individual print and electronic purchase. Briefs are characterized by fast, global electronic dissemination, standard publishing contracts, easy-to-use manuscript preparation and formatting guidelines, and expedited production schedules.

Both solicited and unsolicited manuscripts are considered for publication in this series.

More information about this series at http://www.springer.com/series/8897

Philip Ringrose

How to Store CO$_2$ Underground: Insights from early-mover CCS Projects

 Springer

Philip Ringrose
Norwegian University of Science
and Technology
Trondheim, Norway

Equinor Research Centre
Trondheim, Norway

School of Geosciences
The University of Edinburgh
Edinburgh, Scotland

ISSN 2191-5369   ISSN 2191-5377   (electronic)
SpringerBriefs in Earth Sciences
ISBN 978-3-030-33112-2   ISBN 978-3-030-33113-9   (eBook)
https://doi.org/10.1007/978-3-030-33113-9

This Springer imprint is published by the registered company Springer Nature Switzerland AG
The registered company address is: Gewerbestrasse 11, 6330 Cham, Switzerland

Foreword

Carbon dioxide capture and storage (CCS) is a crucial greenhouse gas mitigation technology. Not only can it decarbonise electricity generation from fossil fuels, where we also have other options, but it can decarbonise industrial sectors where there is no other low-carbon alternative, such as iron and steel production, cement production, refineries and other chemical industries. Even more than that, it can scrub carbon dioxide from the atmosphere, a 'negative emission' technology. These ideas are what lies behind the many analyses and scenarios developed by the International Energy Agency and the Intergovernmental Panel on Climate Change (IPCC) showing that we really need this CCS technology to achieve the 2-degree goal and even more vital for achieving the 1.5 degree ambition. Without CCS, these objectives are likely impossible to meet, and if achieved then at a much higher cost (e.g. 138% higher).

So, it gives me great pleasure in writing a recommendation for this book because it brings together two very important aspects—theory of CO_2 geological storage and learnings from real projects. The theory comes from what has been researched, modelled, published and taught at NTNU and other universities. But the theory only takes us so far. The gems are the learnings from real experience in the field, with suitable reflection and analysis to test theories and to extract further understandings and improve further the theory. Another key aspect is the transfer of this knowledge to others, especially students, who I am afraid to say will have to help deliver far more CCS than we have been able to do so far. In my role at IEAGHG, we try and assist the next CCS generation by running International CCS Summer Schools, we now have completed 13 and have some 600 alumni coming from over 49 countries. There is no question over the enthusiasm and capabilities of this next generation for this challenge. The author of this book has lectured and mentored at several IEAGHG Summer Schools, so I have seen first-hand how good he is at communicating these, at times complex, topics in a clear and easier-to-understand way to students from a wide spectrum of background. This book draws upon all of this. I recommend it.

Tim Dixon
General Manager, IEA Greenhouse
Gas R&D Programme
Cheltenham, UK

Acknowledgements

This short introduction to CO_2 storage could not have been possible without cooperation with numerous people—too many to mention. Industrial-scale projects, like the pioneering Sleipner CCS project, involve hundreds of engineers, technology specialists, managers and suppliers at each stage of the project. It has been my privilege to be a small part of such teams on a few of these projects. The experience I have gained working between the industrial and academic worlds has been especially valuable, and I wish to thank my colleagues, past and present, at Equinor (formerly Statoil), at the Norwegian University of Science and Technology (NTNU), at Heriot-Watt University and at the University of Edinburgh.

Where ever possible I have tried to attribute published sources, but some ideas just permeate. In writing this book, I have certainly borrowed concepts and ideas from colleagues without crediting them properly. So, I take this chance to give special thanks for the 'permeating of ideas and insights' which I know I gained from Anne-Kari Furre, Bamshad Nazarian, Gelein de Koeijer, Michael Drescher, Britta Paasch, Peter Zweigel, Allard Martinius and Guillaume Lescoffit (all at Equinor), Ola Eiken (Quad and Statoil), Martin Landrø (NTNU), Mark Bentley (Tracs and Heriot-Watt), Stuart Haszeldine (University of Edinburgh), Allan Mathieson (BP and Lloyd's Register), Gillian Pickup (Heriot-Watt), Sally Benson (Stanford University), Tip Meckel (BEG, Univ. Texas, Austin) and Tim Dixon (IEAGHG). Martin Landrø took the initiative to establish my position in CO_2 storage at the Norwegian University of Science and Technology, and together we started the first Bachelor/Master module on this topic in 2013. I would especially like to thank the first generations of students who asked tricky questions and helped identify things that are important to know and explain.

The CO_2 storage projects referred to in this book are biased towards Norwegian and European projects and Equinor's CCS operations—so I do apologise for neglecting important experience from elsewhere. Summarising complex topics can be tricky—so if I have over-simplified a topic that you know well, please be provoked to make things clearer for others, and feel free to disagree with my simplifications. Finally, I thank my family (Priscilla, Christy, Juliette, Miriam and

Daniel) for being patient with me for the time and energy I have devoted to CO_2 and rocks. There are other things in life that matter too—but climate and energy are pretty high on the list.

Contents

Nomenclature and Units

The SI International System of Units (Système Internationale d'Unités in French) is used through this book, often also referred to as the metric system of units. However, there are a few exceptions where non-SI units are used for convenience and following common practice:

- Rock unit permeability is usually given in millidarcy (mD). 1 darcy is approximately equal to 1 μm^2 (or 10^{-12} m^2) in SI units.
- Reservoir or well pressure is usually discussed in bars, where 1 bar = 10,0000 Pascals (Pa). Note the difference between bar_g (Gauge Pressure) and bar_a (Absolute Pressure), where bar_g is a reading relative to atmospheric pressure at the location, and bar_a is a pressure reading relative to an absolute vacuum.

A brief description of the main symbols used in this book is given in the table below.

A	Albedo
B_o	Bond number
B_{HC}	Hydrocarbon volume factor
C, C_o	Concentration, reference concentration
C_c	Capacity coefficient
C_a	Capillary number
c	Compressibility
Fm	Formation (stratigraphic)
g	Acceleration due to gravity (on Earth)
h	Height
k, k_r	Permeability, relative permeability
L	Length
P, p	Pressure
P_c	Capillary pressure
P_t	Threshold pressure
Q, q	Flow rate (volumetric)

R_f	Recovery factor
r	Radius
S	Saturation (also solar constant)
T, T_{eq}	Temperature, equilibrium temperature
t	Time
u_x	Flow velocity (in x-direction)
V	Volume
x,y	Horizontal coordinates
z	Vertical coordinate
α	Arrhenius constant
γ	Interfacial tension
Δ, δ	Difference in a property
∇	Gradient of a vector field
ε	Efficiency factor
θ	Angle
λ	Mobility
μ	Viscosity
ρ	Density
σ	Stress (or Stefan-Boltzmann constant)
ϕ	Porosity
Ψ	Ratio

Chapter 1
Why We Need Engineered Geological Storage of CO_2

1.1 Motivation

Reduction in global greenhouse gas emissions is a key issue for modern human civilization. Part of the solution to this challenge is long-term storage of CO_2 in deep geological rock formations. Other key solutions to achieving reductions in greenhouse gas emissions are to greatly expand the use of renewable sources of energy and to use energy much more efficiently. The main purpose of this book is to explain the concepts and technologies involved in the geological storage of CO_2. The material presented here was initially developed as course notes for the Masters' course module entitled 'Operation and Integrity of Engineered Geological Storage of CO_2' at the Norwegian University of Science and Technology (NTNU). The content has also proven useful in several short courses, such as the IEAGHG summer schools and industry courses for experienced professionals 'transferring into' CO_2 storage as a new emerging technology. As you will see this book offers only a short introduction to an extensive and multi-disciplinary topic, where detail and precision are important. However, the reader should not lose track of the essential message: that engineered geological storage of CO_2 is a relatively straightforward and established technology which will be urgently needed in the coming decades. Before digging into the topic of CO_2 storage, it is worth spending some time understanding why we need CO_2 capture and storage (CCS).

1.2 Brief History of Fossil Fuels

Since around 1800, humans have been rapidly increasing their consumption of fossil fuels, starting with coal during the industrial revolution, and then adding petroleum liquids and hydrocarbon gases leading to a significant increase in the rate of fossil-fuel consumption after 1950. This dramatic growth in the use of fossil fuel as an energy source has resulted in the consumption of a very significant fraction (roughly

P. Ringrose, *How to Store CO2 Underground: Insights from early-mover CCS Projects*, SpringerBriefs in Earth Sciences, https://doi.org/10.1007/978-3-030-33113-9_1

one third) of fossils fuel reserves, a resource which accumulated over 0.5 billion years of the Earth's history—coals and petroleum liquids having being derived from the remains of land plants and marine algae deposited and buried during a period of 540 Million Years (Ma).

Figure 1.1 shows historical data on global CO$_2$ emissions compared with some possible future trends. During the industrial and petroleum ages, human society became more and more dependent on increasing levels of fossil fuel combustion, resulting in an acceleration of CO$_2$ emissions to atmosphere. There is now widespread agreement that we need to change this behaviour—we must rapidly reduce this rate of CO$_2$ emissions to atmosphere. Over the last decade, despite many calls to reduce CO$_2$ emissions, all that has been achieved is a slowing in the rate of increase. The 2015 Paris agreement embodies the ambition to dramatically reduce emissions, with the goal of achieving "a balance between anthropogenic emissions by sources and removals by sinks in the second half of this century" (Paris Agreement, COP21, Article 4). To achieve this, a reduction in emissions of more than 50% is needed by 2050, with more detailed assessments of actions needed to avoid a global warming of 1.5 °C requiring pathways reaching around 80% reduction by 2050 (IPCC 2018).

Efforts to achieve this change in behaviour, with regard to greenhouse gas emissions, are generally summarized in terms of an 'energy transition' towards a set of low-carbon or 'green' forms of energy. The most widely accepted model for achieving this transition is the wedge model (Pacala and Socolow 2004), whereby gradual phasing in of renewable energy sources, adoption of energy efficiency measures and application of emissions reduction technologies for fossil fuels could enable this energy transition to occur within around 50 years. To achieve that transition, human societies will need to change their behavioural patterns and adopt new technologies for power generation, transport and industrial activities. At this point in time, most

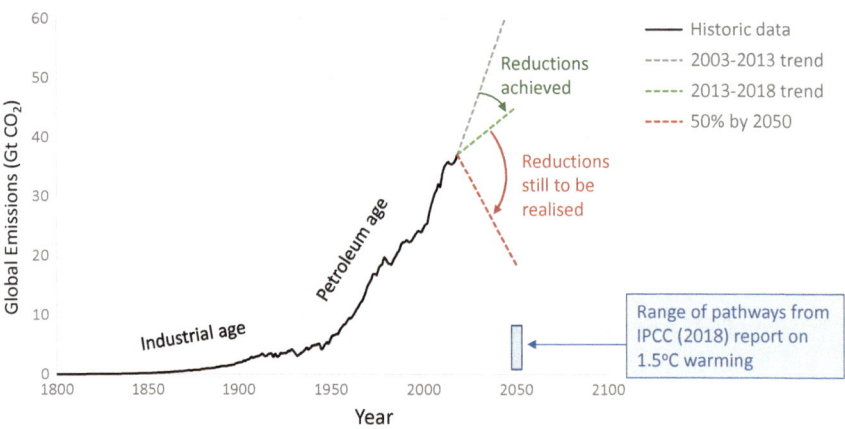

Fig. 1.1 Historical record of global CO$_2$ emissions compared with various projections (data sources: carbon emissions data up to 2013 from https://cdiac.ess-dive.lbl.gov/ with 2014–2018 years estimates from www.wri.org). Figure modified from Stephenson et al. (2019)

societies have already started their energy transition process—however, the speed of change is much too slow.

Note that 'Carbon emissions' and 'low Carbon energy' are popular short-hand phrases for 'CO_2 emissions' and 'low greenhouse gas emissions' (dominated by CO_2). Other greenhouse gas emissions, most notably methane, are generally included as CO_2-equivalent contributions to greenhouse gas emissions, where the term CO_2e is often used.

How we might achieve this energy transition is a highly complex problem which has been much debated. Useful reviews of the problem are given by Stern (2007), Grubb (2014), Stocker (2014), Sachs (2015), and Stoknes (2015). The need for change is widely agreed upon, but whether human society can achieve the transition is a matter of speculation or conjecture. Nevertheless, this complex problem can be more simply captured in terms of the motivation for change and action. Avoiding the harmful effects of global warming is a primary motivation, and the growing concerns about the effects of human-induced climate change are very well captured by the IPCC working groups (e.g. Stocker 2014) as well as a growing set of recent observations on climate change impacts already underway. However, at a more basic level, human-induced climate change is related to an even more fundamental issue— the need to protect our atmosphere. To appreciate the urgent need to protect our atmosphere, it is helpful to review the history of the discovery of the greenhouse gas effect.

1.3 Brief History of Greenhouse Gas

It was the French mathematician Joseph Fourier who first identified (in 1824) that the Earth's atmosphere acts as an insulator, generating a warmer Earth surface than can be explained by solar radiation alone. Life as we know it could not exist without this insulating effect of our atmosphere. These concepts were subsequently developed and formalized in what we now refer to as the Stefan–Boltzmann law of black body radiation, whereby the energy radiated by a 'black body' is proportional to the fourth power of the temperature of that body. The Stefan–Boltzmann law can be re-arranged to give the equilibrium temperature, T_{eq}, for a planet in our solar system:

$$T_{eq} = \left[\frac{S(1-A)}{4\sigma} \right]^{1/4} \tag{1.1}$$

where, S is the solar constant (the intensity of heat radiating from our Sun), A is the bond albedo (a measure of reflectivity) and σ is the Stefan-Boltzmann constant. Assuming reasonable values for $S = 1366$ W/m^2 and $A = 0.3$ (Pollack 1979), gives an equilibrium Earth surface temperature of close to 255 K ($-18\,°C$). The difference between this value and the observed average surface temperature of the Earth, at around 288 K (15 °C), gives us an atmospheric greenhouse warming effect of about

33 °C. By comparing this insulating effect of the Earth's atmosphere with the much smaller effect of atmospheric warming on Mars and the much larger effect on Venus (Pollack 1979) we can also demonstrate the underlying importance of the Earth's atmosphere to life on Earth. Biological systems on Earth have been adapted to life on this rather unique planet, and the atmosphere plays a crucial role in determining the conditions for that biosphere.

In 1896, the Swedish chemist Svante Arrhenius published a paper which was the first study to quantify what we now call the 'greenhouse effect'. By using infrared radiation observations of the moon, Arrhenius was able to calculate the absorption of infrared radiation by atmospheric CO_2 and water vapour (using the Stefan–Boltzmann law). He showed that temperature increase is proportional to the log of CO_2 concentration. Expressing this mathematically we obtain the Arrhenius greenhouse gas absorption law:

$$\Delta F = \alpha \ln(C/C_o) \tag{1.2}$$

where ΔF is the radiative forcing (Wm^{-2}), α is the Arrhenius constant, C is the CO_2 concentration and C_o is a baseline (or unperturbed) concentration of CO_2. Assuming a typical range for α [5.3, 6.3], the radiative forcing effect of increasing CO_2 concentration from the pre-industrial level of 280 ppm to the present level of over 400 ppm is in the range of 1.8–2.2 Wm^{-2}.

Arrhenius' insight was remarkable for several reasons. He was able to quantify the effect of selective heat absorption in the atmosphere due to H_2O and CO_2 by ingenious use of astronomical data—removing the effects of seasonal temperature variations in order to resolve the effect of CO_2. His analysis also set the effects of different atmospheric gases in the context of the solar system as a whole. Our understanding of the greenhouse gas effect has progressed considerably since this early insight, and current work on atmospheric science includes the effects of radiation and absorption at different levels in the atmosphere and the feedback mechanisms associated with ocean circulation and natural carbon sinks and sources. We now know that other trace gases (mainly methane, nitrous oxide, chlorofluorocarbons and ozone) also have strong greenhouse-gas effects. A more recent estimate of the total anthropogenic greenhouse gas forcing effect, ΔF, for the period 1750–2011 is 2.29 Wm^{-2}, with an uncertainty interval of [1.13, 3.33] Wm^{-2} (IPCC 2013).

During the 20th Century, this early appreciation of the greenhouse gas effect, and the critical role of certain greenhouse gas molecules, gradually gained momentum. Developments in methods for measuring atmospheric compositions, either directly or via the gases trapped in ice core data, led to improved models of the history of the (geologically-recent) ice ages, and of the many factors which control on the Earth's climate. Effects of man-made greenhouse gas emissions need to be compared to many other effects, including solar system controls on the earth's orbital patterns, variations in solar intensity, the effects of ocean circulation patterns and volcanic eruptions.

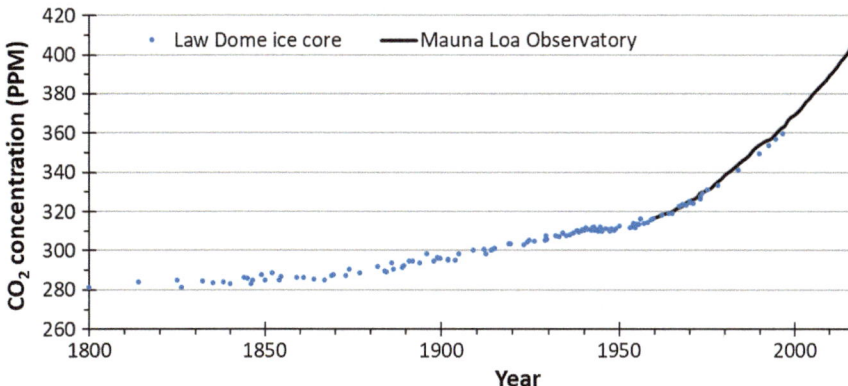

Fig. 1.2 Mean annual CO_2 concentration in the atmosphere from two sources: The Law Dome ice-core dataset (Etheridge et al. 1996; MacFarling Meure et al. 2006); Mauna Loa Observatory measurements from the Earth System Research Laboratory. *Source* www.esrl.noaa.gov/gmd/ccgg/trends/data.html

The importance of CO_2 as a critical greenhouse gas came into renewed focus following the initiative of Keeling and others (Keeling 1978) who started collecting atmospheric CO_2 concentration data at the Mauna Loa observatory in 1958. Figure 1.2 shows this dataset (known as the Keeling curve) alongside CO_2 concentrations based on ice-core data (the Law dome ice core dataset). These data demonstrate a very significant change in atmospheric CO_2 concentration corresponding with the anthropogenic emissions of CO_2 during the industrial age (Fig. 1.1). So far, we have increased the CO_2 concentration in the atmosphere by nearly 50% compared to pre-industrial levels.

This brief review demonstrates several important points which should motivate us concerning the urgent need to achieve the low-carbon energy transition:

1. After many centuries of human society 'taking the atmosphere for granted', pioneering scientists only began to appreciate the critical nature of the greenhouse effect in the last 200 years.
2. After initially benefitting from relatively cheap energy from the burning of fossil fuels (coal, oil and gas), human society started to realise the consequences of the resulting CO_2 emissions over the last few decades.
3. The urgent need to change our habit of uncontrolled combustion of fossil fuels to the atmosphere, is as much about protecting our atmosphere as it is about avoiding the effects of man-made climate change (although the latter is a most serious consequence of the former).

1.4   Why Do We Need CCS?

So why can't we just stop using fossil fuels and switch immediately to renewable energy? The long-term strategy is indeed to achieve a situation with near-zero net emissions to atmosphere, and renewable energy will be key to achieving that objective. However, with current global fossil fuel consumption at around 80% of the world's energy supply (IEA 2016), and with transport, manufacturing and agricultural sectors all being highly dependent on fossil fuel energy, a transition period is needed. Another key challenge for renewable energy sources is their intermittent nature (for wind and solar sources), such that significant growth is only likely to be achieved by using energy storage or by using mixed energy systems (including methane and hydrogen combustion).

Given these challenges for achieving this transition to a low-Carbon energy mix, a more pragmatic approach is needed. This is usually presented in terms of the 'wedge model approach' (first proposed by Pacala and Socolow 2004), in which several parallel activities work together to achieve a significant cumulative effect. There are many versions of the wedge model with varying numbers of wedges, but the dominant components involve achieving global emissions reductions by:

1. Improved energy efficiency measures,
2. Addition of new sources of renewable energy,
3. Switching from coal-fired power to natural gas combustion,
4. Deploying CO$_2$ capture and storage (CCS),
5. Expanded use of nuclear power.

The likely level of growth of these complementary energy options is widely debated, with the IEA and many others presenting annual reviews (e.g. IEA 2015). We focus here on the arguments concerning the use of CCS.

The dominant arguments as to why CCS is vital for achieving the overall greenhouse gas reduction objectives are:

- CCS provides a mechanism for decarbonising both existing power supply and reducing emissions from industry (e.g. cement and steel manufacture).
- CCS allows the energy transition to be achieved faster and at a cheaper cost than by using only renewable energy sources.
- CCS deployed with bio-energy combustion allows negative net-CO$_2$ emissions projects to be deployed.

The important role of CCS has been discussed in several more in-depth analyses of future greenhouse gas emissions scenarios (e.g. IPCC 2013; IEA 2016; Peters et al. 2017; Ringrose 2017), and most recently in the IPCC Special report on Global Warming of 1.5 °C (IPCC 2018). It is also important to note there are various arguments made against CCS. These tend to group into two categories: (i) that CCS will encourage fossil fuel usage to continue longer than necessary, (ii) that CCS is too expensive. Whatever your views on this debate, nearly all forecasts see the need for some level CCS to be deployed—it is just a question of how much CCS will actually be used during the energy transition.

Currently, there are 19 large-scale CCS facilities in operation together with a further 4 under construction, which together have an installed average capture capacity of 36 Mtpa (GCCSI 2019). Considering CCS projects currently in the planning stages, this capacity could increase to 100 Mtpa by around 2030. However, at least a 10-fold increase beyond this level would be needed if the greenhouse gas reduction goals implied by the Paris agreement (COP21 2015) are to be realised. Whatever the actual growth rate in CCS projects globally, the underlying concept is that the captured CO_2 volumes must be stored underground and isolated from the Earth's atmosphere. CO_2 storage technology is therefore vital and will be the focus of the remaining discussion in this book.

1.5 Introduction to CCS Technology

CO_2 Capture and Storage (CCS) refers to a set of technical solutions to remove CO_2 from industrial processes and to inject it into the subsurface in order to isolate the CO_2 from the atmosphere. Note that the terms 'Carbon Capture and Storage' and 'Carbon Sequestration' are also used as short-hand synonyms. CO_2 *capture* mainly refers to processes for removing carbon dioxide from point sources of gas emissions, such as power plants, gas processing facilities and other industrial facilities (especially steel and cement manufacture). CO_2 capture from mobile sources (such as vehicles) and directly from the atmosphere (air capture) is also possible, but both are minor in terms of the volumes of avoided greenhouse gas emissions. There are however growing calls for an expansion in air capture technology and CO_2 capture from biofuels, so this aspect could become more significant in the future.

CO_2 *Storage* means long-term geological storage so as to isolate the CO_2 from the atmosphere for 1000's of years. This process is not usually referred to as permanent disposal for several reasons:

- CO_2 is not a simple waste product (it is also an essential part of the carbon cycle);
- Undesirable emissions of CO_2 to the atmosphere only need, in principle, to be isolated from the atmosphere for a period of a few thousand years;
- Permanent disposal is very difficult to ensure, while it is possible to demonstrate and verify safe long-term geological storage of CO_2.

So, what are the main technologies involved in CCS? There are three essential steps: capture, transport and storage. Figure 1.3 summarizes the main technologies involved in these three stages of CCS.

As we are mainly concerned here with geological storage of CO_2, we only need to appreciate the essentials of CO_2 capture technology. However, as we will see later, transport and storage are rather closely coupled, and we will need a good appreciation of transport technology alongside storage technology.

CO_2 capture technologies can be grouped in several ways. Firstly, two major classes are pre-combustion and post-combustion capture. The pre-combustion technologies all concern removing CO_2 from gas blends where there is a significant

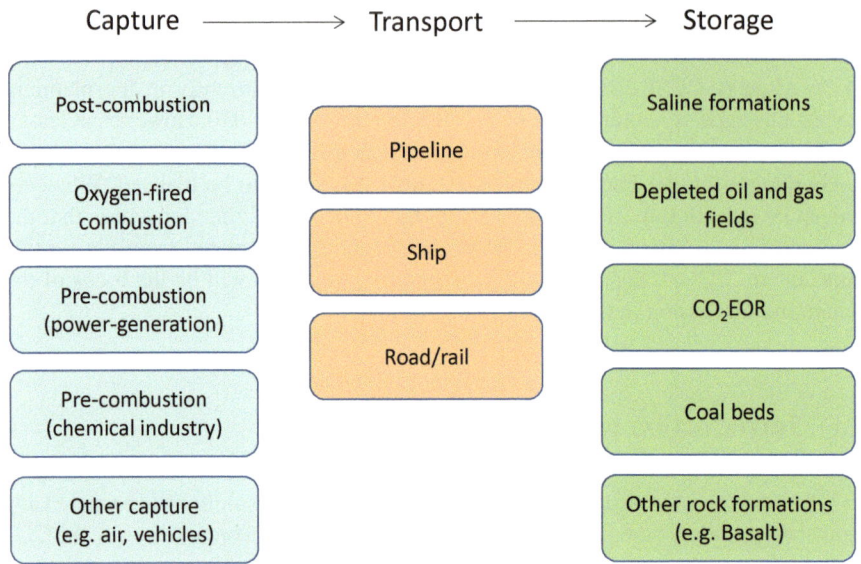

Fig. 1.3 Overview of CCS technologies

fraction of CO$_2$ in the gas. This includes CO$_2$ removal from natural hydrocarbon gas resources as well as CO$_2$ removal from industrial chemical processes which generate CO$_2$ as a by-product (such as fertilizer plants). Post-combustion technologies concern the removal of CO$_2$ from the combustion gas streams (flue gas).

The capture processes themselves can also be grouped by the physio-chemical processes used to separate the gases:

- Solvent-based where an absorption liquid is used;
- Sorbent-based, where solid particles are used;
- Cryogenic, where the different gas condensation temperatures are used;
- Membranes, where solid-state chemical barriers are used.

Another important class of CO$_2$ capture technology is oxygen-fired combustion which is usually combined with air separation and a de-nitrogenation process, where nitrogen is removed from the air before combustion to produce a concentrated stream of carbon dioxide.

Each of these classes have many different technical methods, and each of the major classes can be used in both the pre- and post-combustion processes, so the list of capture technologies becomes quite complex. Herzog et al. (1997), Feron and Hendriks (2005) and MacDowell et al. (2010) give fuller reviews of the main CO$_2$ capture technologies.

So far, the dominant commercial CO$_2$ capture process has been solvent-based using mono-ethanol-amine (MEA) as the solvent. MEA solvents were developed over 70 years ago but have since been modified several times into different variants, to incorporate various inhibitors to resist solvent degradation and equipment corrosion,

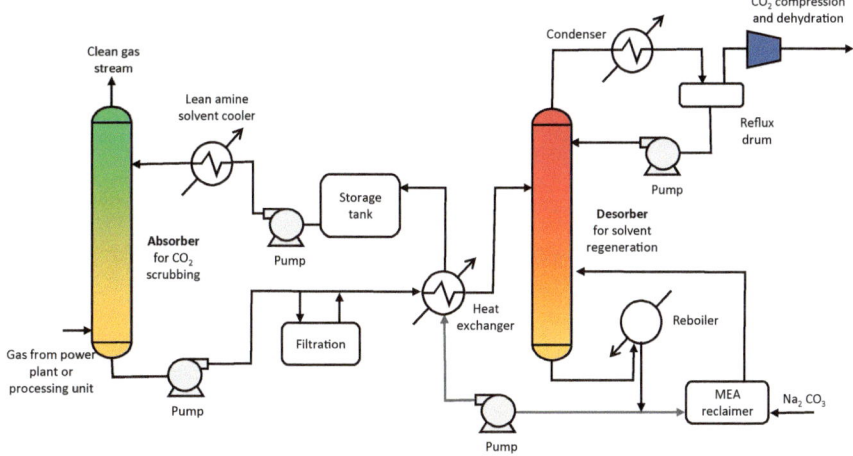

Fig. 1.4 CO_2 capture by the amine separation process (based on Herzog et al. 1997)

and to improve the overall capture efficiency. Figure 1.4 shows the main elements of the amine separation process. Two chemical process columns are used: the absorber to extract the CO_2 from the gas and the desorber to extract the CO_2 from the solvent. Other elements of the process handle the re-generation of the solvent, pumps, heat exchangers, filtration, condensation and gas compression. The process is quite complex and is fairly energy intensive (e.g. heating and compression). Research and development of amine-based CO_2 capture processes has focused on making the CO_2 absorption more efficient, reducing the energy demand, and obtaining more effective recycling of the amine solution. Alternatives to amine-based capture solutions (solid particle sorbents, cryogenic processes and membranes) have great potential in the longer term, but so far post-combustion amine-based solutions still dominate chosen solutions for industrial capture plants, with oxygen-fired combustion and separation also widely applied.

From the point of view of CO_2 storage, we are principally interested in the output from the capture plant. What is the composition, pressure, temperature, and flow rate of the CO_2-rich gas stream, and what is the regularity by which it will be supplied via the transport system?

CO_2 transport technology essentially involves handling the captured CO_2-rich gas/liquid streams and transporting them to the storage site, using pipelines, ships or tankers. This might appear to be simple, but several factors make CO_2 transport more challenging than other gas or liquid transport systems:

- CO_2 has thermodynamic properties which means that it must be handled across the phase transitions—transport of gas-phase, liquid-phase and dense-phase (supercritical) are all likely to occur in CCS systems;
- CO_2 mixed with any aqueous phase becomes corrosive;

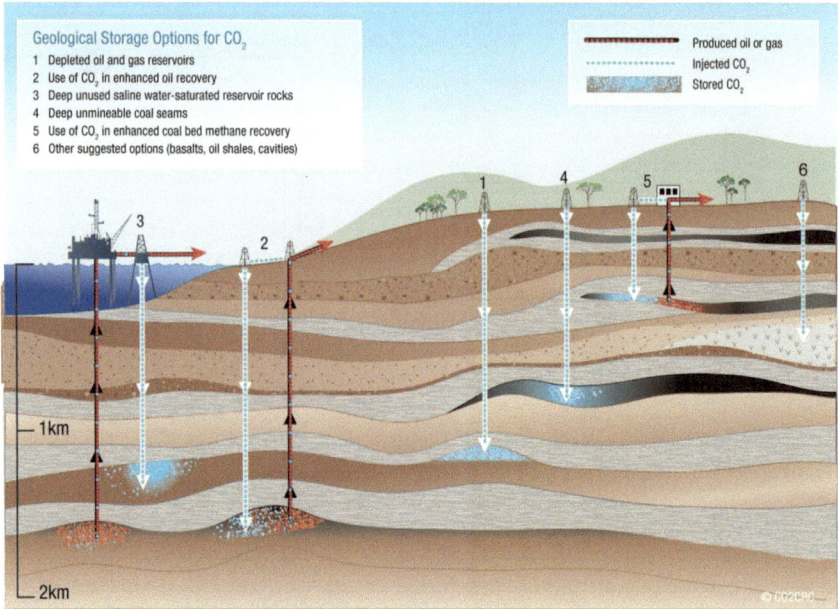

Fig. 1.5 Overview of geological storage options for CO$_2$ (©CO2CRC, image courtesy of CO2CRC Ltd)

- CO$_2$-rich streams contain various other gas components (mainly hydrocarbons, nitrogen, oxygen) which complicate CO$_2$ management;
- CO$_2$ transport and flow assurance technologies are still relatively immature.

To complete the CCS overview, we need to summarise the CO$_2$ storage options. Five main groups of storage options were listed in Fig. 1.3 and these are illustrated in Fig. 1.5.

We summarize these options broadly in terms of their relative potential for storing large volumes of CO$_2$:

1. *Saline aquifer formations*: The largest available volumes for storage of CO$_2$ are found in deep water-bearing reservoirs, also known as saline aquifers. These should not be confused with shallow drinking-water aquifers, hence the term 'deep saline reservoir formations' is often preferred (although 'saline aquifers' is widely used as a summary term).
2. *Depleted oil and gas reservoirs*: This storage option offers great potential in the longer term, because once oil and gas resources have been extracted, these well-characterised reservoirs offer an excellent CO$_2$ storage solution with significant infrastructure in place. They may also have depleted fluid pressures creating some operational challenges but also providing benefits for pressure management.
3. *Storage as a part of CO$_2$ EOR projects*: Where CO$_2$ is used for Enhanced Oil Recovery (EOR), the injected CO$_2$ remains in the subsurface as stored CO$_2$.

CO_2 EOR is usually the main component in the Carbon Capture, Utilization and Storage (CCUS) concept where the economical use of CO_2 is used to help facilitate the economics and growth of large-scale CO_2 capture and storage.

4. *Coal bed storage*: Coal formations can be used either to store CO_2 in un-mineable coal seams or as part of enhanced coal-bed methane projects (analogous to CO_2EOR). These options should not be dismissed but are likely to be relatively minor in terms of the available volume.

5. *Other rock formations*: There are many other suggested options for geological CO_2 storage including volcanic rocks (especially basalt), underground caverns, and oil or gas bearing shale formations. Some of these solutions also involve using mineral reactions to permanently bond CO_2 in a solid form; for example, magnesium silicate rocks (such as serpentinite) react with CO_2 and H_2O to form carbonate minerals. CO_2-hydrates (clathrate hydrate compounds stable at low temperatures) have also been proposed as a potential long-term CO_2 storage option (Zatsepina and Pooladi-Darvish 2012).

The relative merits of these different storage options and the practical steps involved in successfully implementing them are covered in the sections below. For further introductory reading on CO_2 storage concepts refer to Holloway (1997), Benson and Surles (2006) and Cooper et al. (2009).

It is also worth emphasizing that other natural CO_2 storage mechanisms (ocean storage, biosphere storage, and mineral storage), while clearly important in the Earth's carbon cycle, are not the main practical choices for engineered CO_2 storage (mainly because they are too slow). There is wide agreement (e.g. Benson and Surles 2006) that CO_2 storage in deep geological formations provides the main practical solution for the control and reduction of greenhouse gas emissions. Our focus is therefore on CO_2 storage in deep geological rock formations. Undesirable man-made emissions of greenhouse gases need to be compensated by desirable man-made storage of gases that can no longer be emitted to the atmosphere. We call this *engineered geological storage of* CO_2.

References

Benson SM, Surles T (2006) Carbon dioxide capture and storage: an overview with emphasis on capture and storage in deep geological formations. Proc IEEE 94(10):1795–1805

Cooper C (Ed) (2009) A technical basis for carbon dioxide storage: London and New York. Chris Fowler Int 3–20. http://www.CO2captureproject.org/

COP21 (2015) United Nations Climate Change Conference, Paris, France, 30 Nov. to 12 Dec. 2015. 21st yearly session of the Conference of the Parties (COP) to the 1992 United Nations Framework Convention on Climate Change (UNFCCC) and the 11th session of the Meeting of the Parties (CMP) to the 1997 Kyoto Protocol

Etheridge DM, Steele LP, Langenfelds RL, Francey RJ, Barnola J-M, Morgan VI (1996) Natural and anthropogenic changes in atmospheric CO_2 over the last 1000 years from air in Antarctic ice and firn. J Geophys Res 101(D2):4115–4128. https://doi.org/10.1029/95JD03410

Feron PHM, Hendriks CA (2005) CO_2 capture process principles and costs. Oil Gas Sci Technol 60(3):451–459

GCCSI (2019) GCCSI CO2RE database: 2019. Global CCS Institute. https://co2re.co

Grubb M (2014) Planetary economics: energy, climate change and the three domains of sustainable development. Routledge

Herzog H, Drake E, Adams E (1997) CO$_2$ capture, reuse, and storage technologies for mitigating global climate change: a white paper. Massachusetts Institute of Technology Energy Laboratory, Cambridge

Holloway S (1997) An overview of the underground disposal of carbon dioxide. Energy Convers Manag 38:S193–S198

IEA (2015) Carbon capture and storage: the solution for deep emissions reductions. International Energy Agency Publications, Paris

IEA (2016) 20 years of carbon capture and storage: accelerating future deployment. https://www.iea.org/publications

IPCC (2013) Summary for policymakers. In: Climate change 2013: the physical science basis. In: Stocker TF, Qin D, Plattner G-K, Tignor M, Allen SK, Boschung J, Nauels A, Xia Y, Bex V, Midgley PM (eds) Contribution of Working Group I to the fifth assessment report of the intergovernmental panel on climate change. Cambridge University Press, Cambridge, United Kingdom and New York, NY, USA

IPCC (2018) Summary for Policymakers. In: Masson-Delmotte V, Zhai P, Pörtner H-O, Roberts D, Skea J, Shukla PR, Pirani A, Moufouma-Okia W, Péan C, Pidcock R, Connors S, Matthews JBR, Chen Y, Zhou X, Gomis MI, Lonnoy E, Maycock Y, Tignor M, Waterfield T (eds) Global Warming of 1.5 °C. An IPCC Special Report on the impacts of global warming of 1.5 °C above pre-industrial levels and related global greenhouse gas emission pathways, in the context of strengthening the global response to the threat of climate change, sustainable development, and efforts to eradicate poverty. World Meteorological Organization, Geneva, Switzerland, 32 pp

Keeling CD (1978) The influence of Mauna Loa Observatory on the development of atmospheric CO$_2$ research. In: Mauna Loa observatory: a 20th anniversary report. National Oceanic and Atmospheric Administration Special Report, pp 36–54

MacDowell N, Florin N, Buchard A, Hallett J, Galindo A, Jackson G, Adjiman CS, Williams CK, Shah N, Fennell P (2010) An overview of CO$_2$ capture technologies. Energy Environ Sci 3(11):1645–1669

MacFarling Meure C, Etheridge D, Trudinger C, Steele P, Langenfelds R, van Ommen T, Smith A, Elkins J (2006) Law Dome CO$_2$, CH$_4$ and N$_2$O ice core records extended to 2000 years BP. Geophys Res Lett 33:L14810. https://doi.org/10.1029/2006GL026152

Pacala S, Socolow R (2004) Stabilization wedges: solving the climate problem for the next 50 years with current technologies. Science 305(5686):968–972

Peters GP, Andrew RM, Canadell JG, Fuss S, Jackson RB, Korsbakken JI, Le Quéré C, Nakicenovic N (2017) Key indicators to track current progress and future ambition of the Paris Agreement. Nat Clim Change 7(2):118–122

Pollack JB (1979) Climatic change on the terrestrial planets. Icarus 37(3):479–553

Ringrose PS (2017) Principles of sustainability and physics as a basis for the low-carbon energy transition. Pet Geosci 23(3):287–297

Sachs JD (2015) The age of sustainable development. Columbia University Press. ISBN 9780231173148

Stephenson MH, Ringrose P, Geiger S, Bridden M (2019) Geoscience and decarbonisation: current status and future directions. Pet Geosci. https://doi.org/10.1144/petgeo2019-084

Stern N (2007) The economics of climate change: the Stern review. Cambridge University Press

Stocker TF (ed) (2014) Climate change 2013—the physical science basis. Working Group I contribution to the fifth assessment report of the intergovernmental panel on climate change. Cambridge University Press

Stoknes PE (2015) What we think about when we try not to think about global warming: toward a new psychology of climate action. Chelsea Green Publishing

Zatsepina OY, Pooladi-Darvish M (2012) Storage of CO$_2$ as hydrate in depleted gas reservoirs. SPE Reservoir Eval Eng 15(01):98–108

Chapter 2
Geological Storage of CO_2: Processes, Capacity and Constraints

2.1 Introduction

In this chapter we will review the main processes involved in the geological storage of CO_2 and then consider the overall feasibility of storing large volumes of the CO_2 in the deep subsurface. This leads us into an evaluation of the CO_2 storage capacity and the various theoretical and practical constraints for CO_2 storage projects, globally. We will consider the following main topics:

- Overview of the main processes involved in geological storage of CO_2;
- Methods for estimation of CO_2 storage capacity;
- Assessment of the physical limits to storage (including flow dynamics, injectivity and geomechanical constraints).

2.2 Basic Concepts for Geological Storage of CO_2

The basic concept at hand is to store CO_2 captured from sites of anthropogenic CO_2 emissions in underground in rock formations, thus isolating CO_2 from the atmosphere. The rock formations used for long-term storage of CO_2 are porous reservoirs that would otherwise contain water, oil or gas. The two main classes of these porous reservoirs are:

- Saline aquifer formations
- Depleted oil and gas reservoirs.

Less important classes of porous rock formations (by volume) include coal beds, shale units, volcanic rocks and underground caverns. In terms of the key processes involved, we will focus on saline aquifer formations and depleted oil and gas reservoirs. However, many of the processes involved apply equally well to other storage domains, and to storage as part of CO_2 EOR projects.

P. Ringrose, *How to Store CO₂ Underground: Insights from early-mover CCS Projects*, SpringerBriefs in Earth Sciences, https://doi.org/10.1007/978-3-030-33113-9_2

A second underlying concept is that we need to store the CO_2 relatively deep (greater than ~800 m) to ensure that CO_2 is in a dense form—either a liquid or as a super-critical phase (Fig. 2.1). This is important for storage efficiency (a higher density means more effective storage) and leads to the widely accepted principle that CO_2 storage sites need to be deeper than about 800 m (although the actual transition to the liquid phase also depends on temperature and the local geothermal gradient). Depth is also important for storage security, and this leads us to the third essential concept—the need for a storage seal. As we approach depths of around 1 km or more we enter the domain of compacted and cemented rocks which potentially contain low-permeably sealing units (e.g. shales, faults, and salt units). At these depths we know from experience that natural gas has been trapped beneath geological seals for millions of years, and so the potential for long-term trapping of CO_2 at these depths is also clearly possible.

Figure 2.2 shows a conceptual shallow stratigraphic sequence representative of the North Sea basin, illustrating how a deeper CO_2 storage target formation overlain by a mudstone sequence could constitute a CO_2 storage system. The actual properties of sealing units should be determined via site investigation and appraisal studies.

Having established the basic concepts defining the storage target—deep and porous rock formations with sealing potential—we can begin to address the specific questions for identifying suitable storage sites.

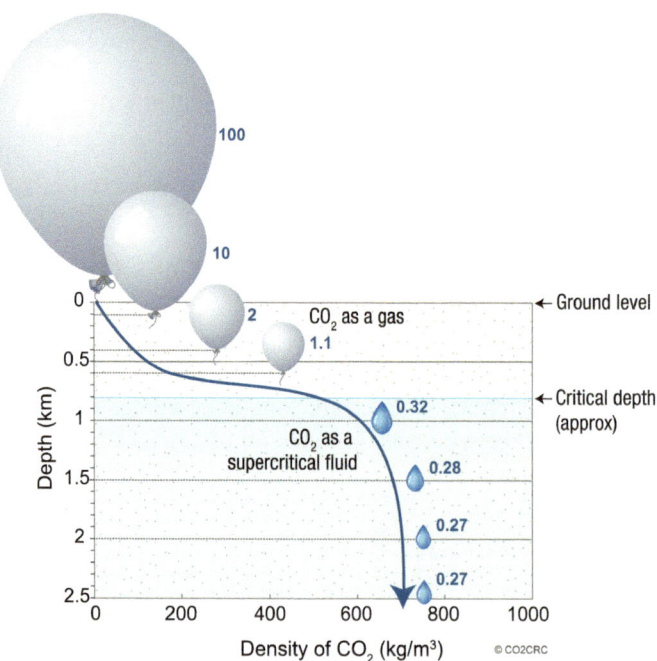

Fig. 2.1 Simplified CO_2 density versus depth diagram (©CO2CRC, image courtesy of CO2CRC Ltd)

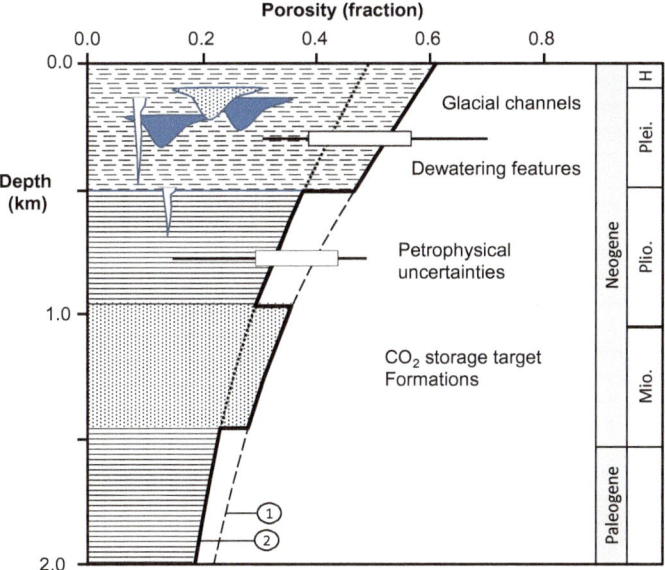

Fig. 2.2 Conceptual sketch showing a shallow stratigraphic sequence representative of the North Sea basin. Typically, a Miocene CO₂ storage target formation would be capped by Pliocene mudstone sequences forming the main containment system. The role of shallow glacial channel and dewatering features in the overlying Pleistocene sequence may be a key issue for assuring storage containment. Reference porosity curves are shown based on (1) Sclater and Christie (1980), and (2) Marcussen et al. (2010). The actual porosity and permeability of the shallow basin sequence (<1000 m) is variable and uncertain and would need to be determined via site investigation studies

The essential questions for any CO₂ storage project are:

- Where can we store the CO₂?
- How much CO₂ can we inject?
- Can we store it safely?
- Can we store it cost-effectively?

These general questions underpin the three main storage issues (Fig. 2.3), and some more project specific questions:

- *Capacity*—is there room for the required CO₂ storage volume over the project lifetime?
- *Injectivity*—will we able to inject the CO₂ at a sufficient rate using the available injection wells?
- *Containment*—will the CO₂ remain in the geological storage unit or could it migrate to another geological formation or even leak out?

We will address these three technical issues in more detail in the following sections. At this point it is useful to link these key questions and issues to the main phases of a CO₂ storage project (Fig. 2.4).

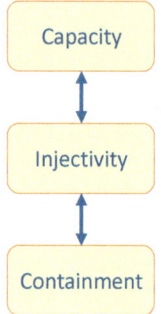

Fig. 2.3 The three key storage issues

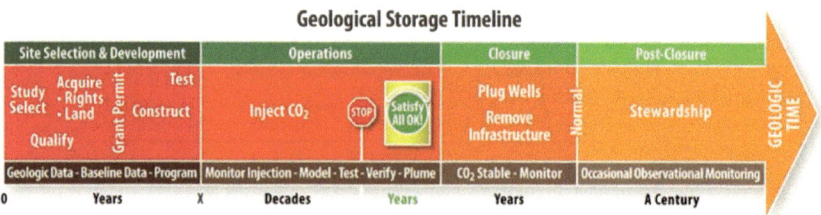

Fig. 2.4 Main phases of a CO_2 storage project (from Cooper et al. 2009; Reproduced with permission, CO_2 capture project, www.CO2captureproject.org/)

1. Site selection and development
2. Site operation
3. Site closure
4. Post-closure stewardship.

The three key storage issues are of course important throughout all the phases, but in terms of focus the capacity issue tends to be foremost in the site selection phase, the injectivity issue dominates in the site operation phase, while containment is the essential question for the site closure and post-closure phases.

There is also a principle of maturity and refinement during the site selection and development phase. As more data is acquired (from regional mapping studies, site surveys, and drilling of exploration and appraisal wells), the uncertainties concerning estimations of capacity, injectivity and containment should gradually decrease towards an acceptable level, when the decisions to proceed with project milestones and eventual site operation are made. If during site development and mapping, any of these issues falls below a certain cut-off level (capacity or injectivity turns out to be smaller than needed, or containment barriers begin to look very weak) then rejection of the site becomes the appropriate action.

Rock systems are inherently complex, and we therefore need a method to define the storage system and the volume around it. It is important to define several key terms describing and encompassing the storage site:

- The sedimentary basin which contains the proposed storage units;
- The storage complex which defines the storage reservoir(s) and sealing units;
- The storage unit(s) themselves, referring to specific geological units;
- The sealing formations and faults;
- Study areas defined for the project or site investigation (e.g. area of review, site boundary, monitoring or survey areas).

In terms of the regulations governing CO_2 storage, the *storage complex* is the key volume to be defined, which under the EU CCS Directive (EC 2009; annex 1) is defined as follows:

Sufficient data shall be accumulated to construct a volumetric and three-dimensional static (3-D)-earth model for the storage site and storage complex, including the caprock, and the surrounding area, including the hydraulically connected areas.

The EU Directive goes on to define 'leakage' as 'any release of CO_2 from the storage complex', while 'significant irregularity' means 'any irregularity in the injection or storage operations or in the condition of the storage complex itself, which implies the risk of a leakage or risk to the environment or human health.' We will revisit these topics when we address the issue of storage integrity and site management (Sect. 3.3). Here we will review the methods for assessing Containment, Capacity and Injectivity. Note that different regions/continents have different regulatory frameworks in place, so we will only use the EU Directive as an example in this discussion.

2.3 Containment and Trapping Mechanisms

2.3.1 Trapping Mechanisms Overview

The containment theme needs to be addressed first and is itself composed of several mechanisms for trapping or retaining CO_2 in the subsurface. There are many ways to group these trapping and containment mechanisms, but most fundamentally they can be grouped into physical and chemical factors:

1. Physical trapping mechanisms related to basin-scale processes, including:

   - Regional structure, basin history, fluid flow and pressure distribution;

2. Physical trapping mechanisms related to the geometry of structural and stratigraphic traps:

   - Controlled by the rock architecture of the storage complex;

3. Physical trapping mechanisms related to fluid flow processes, principally:

    - Capillary interfaces between fluids;
    - Retention of CO_2 as a residual phase;

4. Geochemical trapping mechanisms, principally:

    - CO_2 dissolution in the brine phase;
    - CO_2 precipitation as mineral phases.
    - CO_2 sorption/absorption (e.g. on clay minerals).

These mechanisms were conceptually combined in the IPCC special report (Metz 2005) into a trapping versus time diagram (Fig. 2.5) illustrating the principle that the various trapping mechanisms should work together to increase storage security as a function of time. While there is much debate about the rates and magnitudes of the various trapping mechanisms, there is wide agreement that this concept is qualitatively correct. (From Benson et al. 2005; © Cambridge University Press, reproduced with permission).

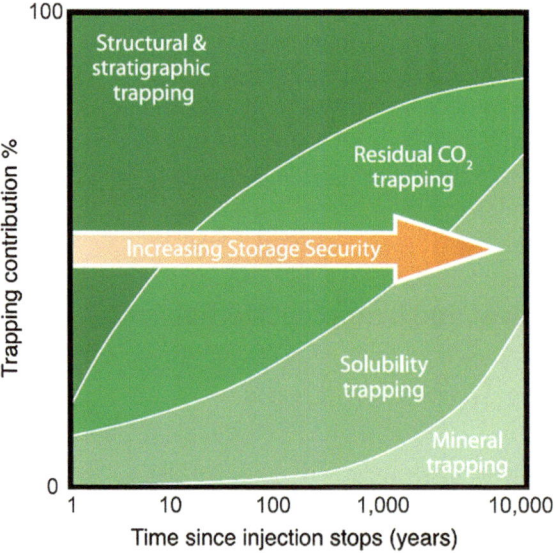

Fig. 2.5 CO_2 storage trapping mechanisms (from Benson et al. 2005, © Cambridge University Press, reproduced with permission)

2.3.2 Capillary Trapping

The first two mechanisms in this diagram depend on the physical process of capillary trapping caused by the interfacial tension at the interface between two fluids in a porous medium (Fig. 2.6).

The role of interfacial tension as the critical phenomenon controlling the size for oil and gas accumulations in the subsurface has been widely demonstrated, and the same approach can be applied to natural or man-made accumulations of CO_2. Berg (1975) defined the thickness of a gas or oil column, z_g, that can be retained against gravity by the capillary entry pressure of the sealing rock as:

$$z_g = \frac{2\gamma \cos\theta \left(1/r_{cap} - 1/r_{res}\right)}{g\left(\rho_w - \rho_g\right)} \tag{2.1}$$

where, r_{cap} and r_{res} are the pore throat radii in the cap rock and reservoir, γ is the interfacial tension, θ is the fluid contact angle and ρ_w and ρ_g are the densities of water and gas. This principle is illustrated by the analytical models shown in (Fig. 2.7), applicable to any buoyant non-wetting fluid.

Consequently, if we know the pore throat radii, the interfacial tension, the fluid contact angle and the fluid densities, we can estimate the maximum CO_2 column height that can be held behind a given caprock unit. Note that the fluid properties are strongly dependent on temperature, pressure and composition (especially the brine salinity and the effect of minor gas components). A fluid contact angle of zero corresponds to perfect wetting behaviour (where the fluid adheres to the rock surface due to molecular forces), while a fluid contact angle of $\theta = 180°$ corresponds to perfectly non-wetting behaviour (the rock surface repels the fluid).

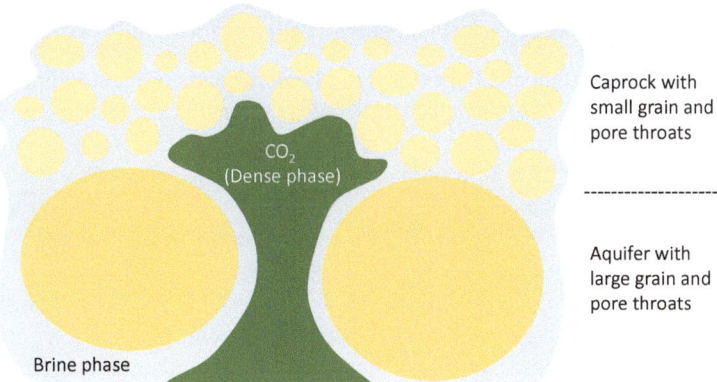

Fig. 2.6 Simple sketch of capillary trapping of a CO_2 phase in a water-wet porous medium at the interface between large and small pore throats at the aquifer-caprock interface

(a)   **(b)**

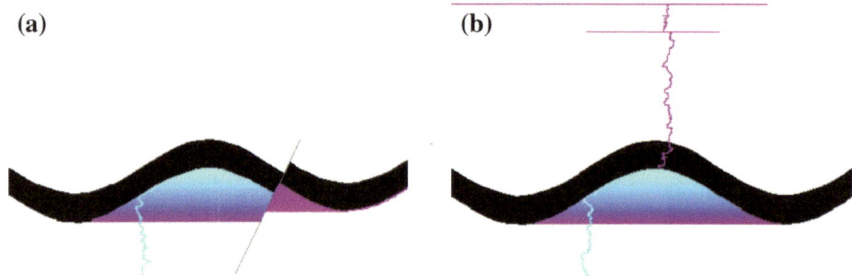

Fig. 2.7 Analytical capillary trapping models (modified from Ringrose et al. 2000): **a** filled trap with leaky fault and tight caprock (leaking via a lateral spill point); **b** filled trap leaking through caprock (due to a lower capillary threshold pressure in the caprock unit). Caprock layer in black; colours denote migration time for the non-wetting phase (blue early; purple late). Analytical model simulated with the invasion percolation tool Permedia-Mpath

Naylor et al. (2011) have compared CO_2 and hydrocarbon column heights, using a compilation of measured interfacial tension data for a range of pressures and temperatures. They proposed a column height ratio, Ψ, in order to compare CO_2 and hydrocarbon gas columns for a given caprock, where:

$$\Psi_{gas/CO_2} = \frac{\Delta\rho_{gas/water}}{\Delta\rho_{CO_2/water}} \frac{\cos\theta_{CO_2/water}}{\cos\theta_{gas/water}} \frac{\gamma_{CO_2/water}}{\gamma_{gas/water}} \qquad (2.2)$$

In general, Naylor et al. (2011) observed that:

- The capillary entry pressure for pure CO_2/water systems is up to 50% lower than for gas/water systems;
- The buoyance force is however lower due the higher density of CO_2;
- These effects tend to cancel each other out so that column heights for CO_2 and CH_4 are about the same (although generally slightly lower for CO_2).

The second important capillary trapping phenomenon is residual trapping within the storage unit itself. This is a well-established factor in limiting oil and gas recovery from hydrocarbon fields, and the same principle applies to CO_2. Taking the case of a migrating plume of CO_2 in a permeable reservoir unit, as the CO_2 rises due to buoyancy, a trail of residual gas/liquid is left behind due to pore-scale trapping of CO_2 (Fig. 2.8). The degree of trapping is controlled by several factors, especially the pore throat size, the interfacial tension and the wettability. While it is generally correct to assume that CO_2 behaves as a non-wetting phase in sandstone reservoirs, some situations may give rise to partial wetting behaviour, especially for carbonate and clay mineral surfaces. The contact angle also varies with pressure, temperature and fluid composition.

Figure 2.9 shows a very simple desk-top experiment to illustrate the nature of pore-scale trapping of a non-wetting phase (in this case, olive oil) within a water-wet porous medium. To measure this effect for liquid/dense phase CO_2 at in situ

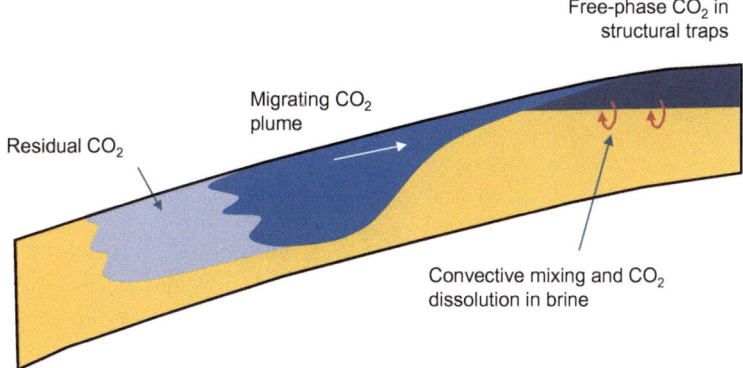

Fig. 2.8 Sketch illustrating CO_2 storage flow processes

conditions is of course more complex, but effect is similar—approximately 20% of the non-wetting phase becomes trapped (the residual saturation).

These pore-scale processes (controlled by molecular phenomena), are summarized in fluid engineering terms as the residual CO_2 saturation. Figure 2.10 shows example CO_2-brine relative permeability curves where both the drainage and imbibition flooding cycles were measured, resulting in a measured residual CO_2 saturation of 22% ($S_w = 0.78$). However, it is not only pore-scale processes that control the residual saturation; it also depends very much on the rock heterogeneity and fluid dynamics, such that the true degree of residual CO_2 trapping is usually calculated using dynamic flow simulations (although analytical approaches can also be used). Furthermore, in order to generate residual trapping, the CO_2 plume must migrate—a gravity-stable body of CO_2 within a structural trap and filling from the top would have essentially no residual trapping volume. Accurate measurement of residual CO_2 trapping is an active field of research, with recent insights and reviews provided by Krevor et al. (2015) and Reynolds and Krevor (2015).

2.3.3 Geological Controls and Site Characterisation

Given a good understanding of the fluid physics governing CO_2 trapping mechanisms, it is important to emphasize (perhaps too obvious for some) that it is the geological system that ultimately controls the effectiveness of CO_2 containment and trapping mechanisms; hence all the efforts required in site assessment and reservoir characterization in any real CO_2 storage project (e.g. Gibson-Poole et al. 2008). Perhaps a little less obvious is how the multi-scale nature of geological systems requires that many different geological processes and phenomena need to be assessed in

Fig. 2.9 Simple visualisation of capillary trapping in a porous medium. In this experiment olive oil is retained in a granular water-wet porous medium (gravel clasts of around 2–3 mm in diameter) with blue dye added to the water phase. Around 20% of the olive oil is prevented from migrating to the top of the sealed column due to capillary trapping in the pore spaces. Note that at room temperature olive oil has a density of around 910 kg/m^3 so the buoyancy force is quite weak compared to CO$_2$ in the subsurface. The interfacial tension for olive-oil/water at room temperature is around 32mN/m (Sahasrabudhe et al. 2017) which is actually quite similar to dense-phase CO$_2$ in the subsurface (Naylor et al. 2011). Vertical scale is in mm

order to evaluate the various trapping mechanisms. This in turn means that geological characterization work must encompass basin scale processes, structural geology, sedimentary geology and small-scale petrophysical analyses (Fig. 2.11).

The work processes for such site characterization work are well established from the petroleum industry based on the experience (and business driver) for oil and gas field developments. However, CO$_2$ storage projects which currently lack financial incentives, may struggle to defend the large up-front investments in site characterization work. On the other hand, CO$_2$ projects may be able to build on previous work in petroleum reservoir characterization in mature sedimentary basins, since many CO$_2$ storage prospects are found in the same basins as oil and gas. Co-location of CO$_2$ storage with oil and gas field developments brings further advantages, including joint use of facilities, exploration wells and seismic data. Figure 2.12 illustrates some

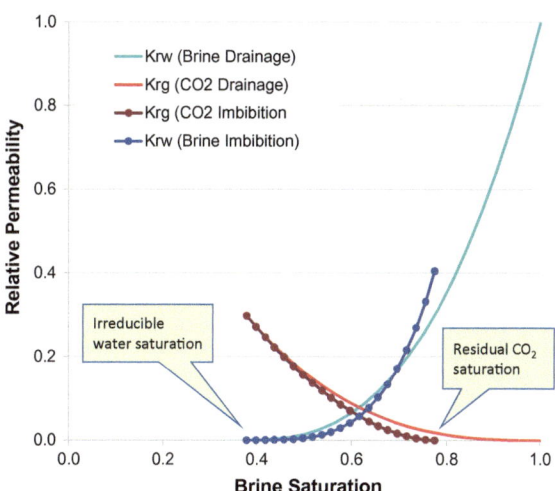

Fig. 2.10 Example CO_2-brine relative permeability curves from Bennion and Bachu (2006) (Cardium Sandstone; IFT = 56.2 mN/m)

examples of site characterization datasets assembled for the In Salah CO_2 storage site (Ringrose et al. 2011). However, while building CO_2 storage site datasets and knowledge based on previous work on petroleum systems is important and useful, it does not remove the underlying need for dedicated site data specifically designed for the CO_2 storage problem (covered by the three main issues: capacity, injectivity and containment).

When conducting site characterization work, we should not lose sight of the simple fact that CO_2 is to be stored inside the pores of underground rock formations. This is sometimes referred to as *pore-space characterisation,* which is a general term to cover a broad set of petrophysical and petrographical methods designed to quantify the nature of rock pores, including basic properties such as porosity and permeability, but also more complex properties such as the mineralogy of pore surfaces, chemical reactivity, pore-throat size distributions and wetting behaviour. Figure 2.13 illustrates examples from pore-scale characterisation work done as part of the In Salah CO_2 storage demonstration project (Lopez et al. 2011). In this example, as well as relating the flow properties (porosity, permeability, CO_2/water relative permeability measurements and endpoint saturations) to different rock types and pore types, the study found that thin pore-coating clay minerals (chlorite) had a significant influence on the macroscopic flow properties.

2.3.4 Geochemical Trapping Processes

CO_2 is a naturally-occurring substance in the subsurface rock system, occurring both as a dissolved component of aqueous fluids (groundwater) and as a free/mobile gas phase. The main sources of this naturally-occurring CO_2 are from (i) volcanic systems

Fig. 2.11 Rock architecture at multiple scales (clockwise from top left): Lamina-scale permeability variations (Tilje Formation, Norway); Normal fault with fault gouge and clay smear (Sinai, Egypt); Tidal delta sedimentary Architecture (Niell Klinter Formation, Greenland); Faulted Devonian siliciclastic sequences (Jameson Land, East Greenland)

with the CO_2 being sourced from the deep mantle (Gilfillan et al. 2008) and from (ii) gas generated from buried organic sources. The major natural accumulations of CO_2 in North America (e.g. Bravo Dome, New Mexico and Sheep Mountain, Colorado), which are used as sources for CO_2EOR projects, contain CO_2 of a predominantly mantle origin. CO_2 is also produced from a wide range of biologically-sourced systems, including decomposition of organic matter, methanogenesis (a by-product of methane producing microbes), oil-field biodegradation, hydrocarbon oxidation, and decarbonation of marine carbonates.

CO_2 is a molecule that is integral to the Earth's carbon cycle, being generated and consumed in a wide range of natural and man-induced chemical reactions. One of the

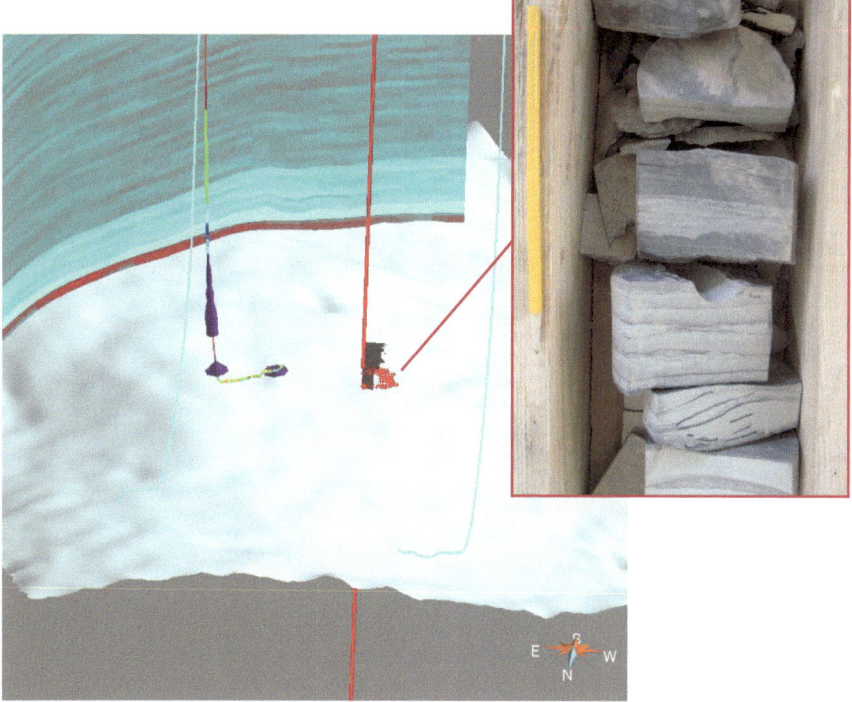

Fig. 2.12 Example of site characterization datasets assembled for the In Salah CO_2 storage site: CO_2 injection wells (blue), appraisal wells with porosity (red), calliper (grey), gamma (colour) logs, pre-injection CO_2 gas distribution (purple); core samples (insert); section shows reservoir and caprock porosity estimated from seismic and well data and surface shows base reservoir mapped from seismic

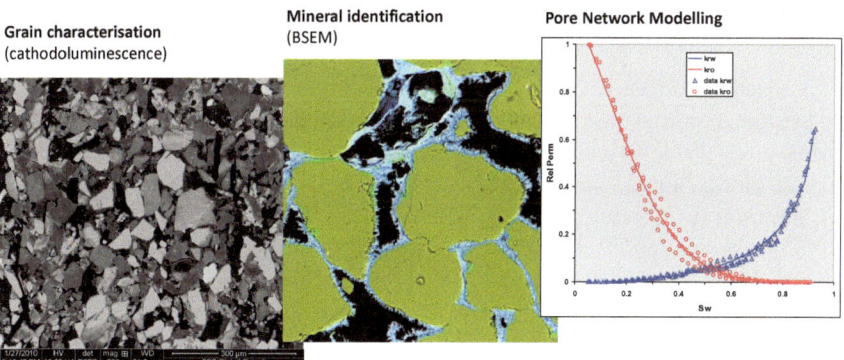

Fig. 2.13 Examples of pore-scale characterisation work done as part of the In Salah CO_2 storage project (Lopez et al. 2011), including petrographic core analysis of grains and pores (using cathode luminescence) and backscatter scanning electron microscopy (BSEM) to identify pore structure and mineralogy, followed by pore-scale modelling to estimate relative permeability curves for different rock types (Images courtesy of Equinor)

most important natural reactions is carbonate dissolution when, for example, shelly material from marine organisms is dissolved in acid (rainwater is slightly acidic), generating CO_2 in the process:

$$CaCO_3(s) + 2\,HCl(aq) \rightarrow CaCl_2(aq) + CO_2(g) + H_2O(l) \qquad (2.3)$$

When surface waters are saturated with carbon dioxide, calcium carbonate will react to form calcium bicarbonate, an important reaction in the weathering of carbonate rock formations, forming limestone caverns and causing scale to precipitate in regions with hard water. The basic carbonate weathering reaction is:

$$CaCO_3 + CO_2 + H_2O \rightarrow Ca(HCO_3)_2 \qquad (2.4)$$

Conversely, carbonate minerals may be formed when calcium hydroxide (Portlandite) reacts with CO_2 (from air or water), an important reaction in concrete and in wellbore cements:

$$Ca(OH)_2 + CO_2 \rightarrow CaCO_3 + H_2O \qquad (2.5)$$

These important reactions might lead one to suppose that injection of CO_2 into the surface would lead to some rather dramatic dissolution and precipitation reactions. However, geological data from natural analogues (e.g. Baines and Worden 2004) demonstrates that:

- When CO_2 is added to pure quartz sandstones, as soon as the formation water is saturated with CO_2, the injected CO_2 will simply remain as a separate phase;
- In the case of CO_2 injection into carbonates (or rocks with carbonate cements), some dissolution of carbonate minerals will occur, but again as soon as the formation water becomes saturated with CO_2, the injected will remain as a separate phase.

Experience from early CO_2 storage injection projects, such as Sleipner, In Salah and Snøhvit, confirms that geochemical reactions are slow and relatively minor (e.g. Carroll et al. 2011; Black et al. 2015), with virtually all the CO_2 remaining as a separate phase (liquid, gas or dense phase). In an analysis of data from a natural CO_2 reservoir (a CO_2-rich gas field), Wilkinson et al. (2009) showed that 70–95% of the CO_2 is present as a free phase, after tens of millions of years, with only around 2.4% of the CO_2 stored in the mineral phase and a similar amount dissolved in the pore waters. The overall point here is that although dissolution and precipitation reactions do occur when new CO_2 is introduced into the subsurface, the CO_2 quickly establishes a new chemical equilibrium with the in situ pore waters, following which reaction rates are very slow. CO_2 dissolution into the brine phase can however be significant (see below).

When CO_2 is put into contact with clay minerals, the possible reactions and effects that can occur become rather complex. In an analysis of CO_2 storage in shales, Busch et al. (2008) showed that gas sorption could lead to significant CO_2 storage

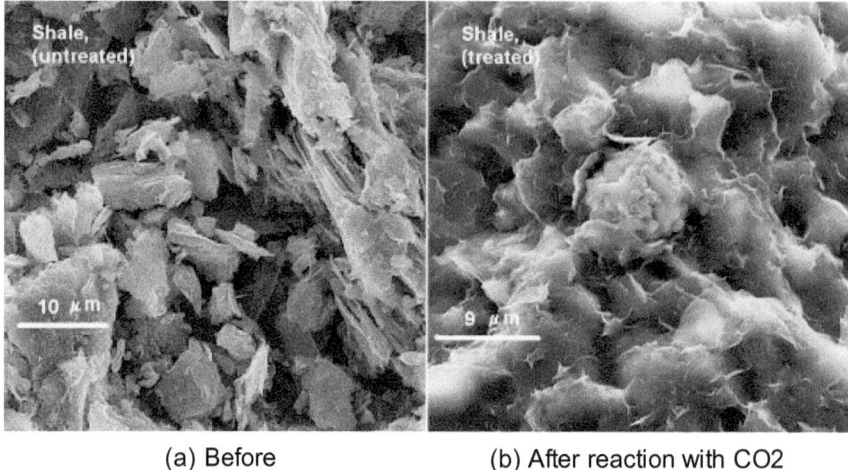

(a) Before   (b) After reaction with CO2

Fig. 2.14 Effect of CO_2 reaction with shale (Kaszuba et al, 2003; © Elsevier, reproduced with permission)

capacity in shale sequences; however, this fixation of CO_2 in the mineral system is related to a combination of processes including aqueous solubility, geochemical reactions and physical sorption on clay minerals (Fig. 2.14). Geochemical reactions such as dissolution of silicate minerals and precipitation of carbonate minerals may potentially have a measurable effect on the porosity, permeability and the diffusion properties of shales.

So then, to summarise the geochemical aspects of CO_2 storage, some trapping (or bonding) of injected CO_2 as a mineral phase can occur, but the rates of reaction are very slow. Some dissolution of carbonate minerals may also occur, but again at very slow rates. Assessment of a potential geological storage site will therefore need some site-specific assessment of these processes, and generally reactions in the wellbore environment are of greatest concern. This aspect will be reviewed below in Chap. 3.

2.3.5 CO_2 Dissolution

The most important geochemical reaction for CO_2 storage projects is CO_2 dissolution in the brine phase. This process has an important potential to assist and stabilise long-term storage but estimates of the effect vary enormously. We know that the process of molecular diffusion of CO_2 within a saline aqueous phase is very slow (see Niemi et al. 2017 for a fuller discussion of this process). We also know that the convective mixing at the CO_2-brine interface is much faster process and so dominates the rate of CO_2 dissolution (Fig. 2.15). To initiate convective mixing, a diffusive boundary layer needs to develop and must achieve a critical thickness before convection can occur.

Fig. 2.15 Density-driven
flow in CO_2 storage in saline
aquifer (from Pau et al. 2010,
© Elsevier, reproduced with
permission)

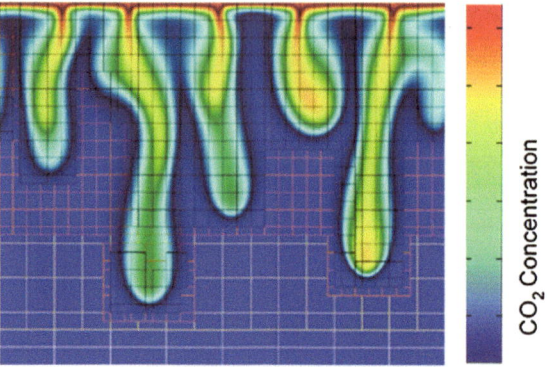

Using numerical analysis based on experimental data, Riaz et al. (2006) showed that
the critical time (t_c) for onset of convection and the characteristic wavelength (λ_c)
are estimated to be in the range of:

- 10 days $< t_c <$ 2000 Years
- 0.3 m $< \lambda_c <$ 200 m.

This is a rather large range but illustrates the expected dimensions and the uncertainties involved. In practice, the in situ fluid conditions and the detailed geological
architecture will determine how fast this process will proceed. For the Sleipner case
where we have good monitoring data on the growth of the plume, it is likely that the
actual in situ CO_2 dissolution rate is between 0.5 and 1% per annum, or around 10%
after 20 years (Alnes et al. 2011; Cavanagh et al. 2015; Ringrose 2018). This degree
of CO_2 dissolution is also consistent with recent laboratory measurements of dissolution rates in porous media (with supercritical CO_2 at in situ conditions) showing a
0.5 m thick CO_2-saturated water layer below the CO_2 layer after 5 years, implying
a 2 m thick layer would be expected at Sleipner after 20 years (Amarasinghe et al.
2019).

2.4 Calculating Storage Capacity

2.4.1 How Much Is a Tonne of CO_2?

Most projects concerned with CO_2 capture and storage talk about Millions of tonnes
(Mt) of CO_2. This contrasts with the hydrocarbon industry which generally refer to
barrels (or standard cubic metres) of oil and billions of cubic feet (bcf) of gas (or
standard cubic metres). So how much is a tonne of CO_2? The essential reason we
work with mass is that CO_2 volume is highly dependent on pressure and temperature.
Mass = volume × density, so at standard surface conditions one tonne of CO_2 has a
volume of 534 m^3; however, at a depth of around 1 km (corresponding approximately

Table 2.1 Comparison of CO_2 density at different standard conditions

Standard reference	Reference pressure	Reference temperature (°C)	Density (kg/m^3)
Chemistry (IUPAC)	1 bar (0.9869 atmosphere)	0	1.976
National Standards (NIST)	1 atmosphere, 1.013 bar	20	1.842
International Standards (ISA and ISO)	1 atmosphere (1.013 bar)	15	1.87
Petroleum Engineers (SPE)	1 bar (0.9869 atmosphere)	15	1.848

to the injection point for the Sleipner project) one tonne of CO_2 has a volume of 1.43 m^3 (assuming a density of 700 kg/m^3).

The precise calculation of the subsurface volume of a tonne of CO_2 is dependent on many factors, especially how well you know the actual in situ pressure and temperature, but also what you assume about the thermodynamic conditions (e.g. assuming isothermal or adiabatic thermodynamic processes) and the composition of the injection fluid (which may contain some methane or nitrogen). Standard practice is to convert injected surface volumes to mass, with reference to known surface conditions or to standard conditions. However, different disciplines use different standards (Table 2.1)—so take care you use a consistent standard.

Most gas engineers work with surface volumes in standard cubic feet (scf) or MMscf (10^6 scf) or Bscf (10^9 scf), while CO_2 projects prefer to report mass (for good reasons). So, some appreciation of the conversion factors is useful. For example, at standard conditions (ISA; 1.013 Bar & 15 °C):

- 1 m^3 of CO_2 has a mass of 1.87 kg
- Since 1 Bscf $= 28.32 \times 10^6$ m^3 the mass of 1 Bscf of $CO_2 = 52{,}959.5$ Tonnes
- Mass of 1 MMscf is then 52.96 Tonnes.

So, a single well injecting 20 MMscf per day is injecting about 1000 tonne of CO_2 per day, and a single well injecting 1 Mt CO_2 per year has a surface injection rate of about 18.8 Bscf per year.

For real injection sites the actual reservoir pressure and temperature is imprecisely known, such that in situ density can only be approximately estimated. Figure 2.16 shows a selection of CO_2 density functions (assuming isothermal conditions) with the shaded regions indicating estimated in situ reservoir conditions for the Sleipner (A), In Salah (B) and Snøhvit (C) storage sites. The largest uncertainty is for the Sleipner case because the shallow depth places it close to the critical point and the phase transition between the vapour and supercritical states. Several studies have used integrated monitoring data and modelling studies to estimate the in situ CO_2 density at the Sleipner site (Bickle et al. 2007; Singh et al. 2010; Alnes et al. 2011; Cavanagh and Haszeldine 2014). At the point of injection into the reservoir the CO_2 has a density of around 485 kg/m^3, however due to cooling into the formation the density is expected to increase away from the well. Using gravity field monitoring observations, Alnes et al. (2011) estimated the average CO_2 density in the plume to

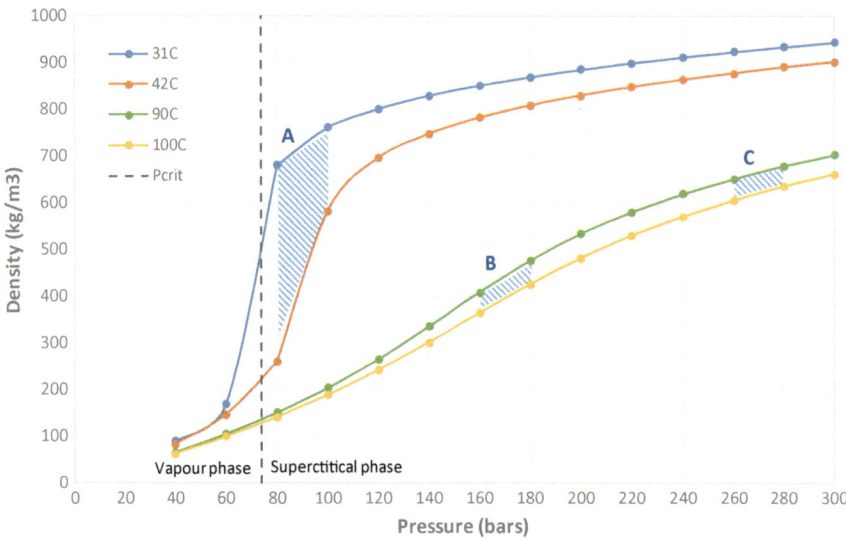

Fig. 2.16 CO₂ density functions for selected isothermal temperatures (NIST Chemistry WebBook). Shaded regions indicate estimated in situ reservoir conditions for the Sleipner (A), In Salah (B) and Snøhvit (C) storage sites

be 675 ± 20 kg/m³. However, assuming slightly higher temperatures in the reservoir gives lower density values. For example, using a modelling approach, Cavanagh et al. (2015) estimated densities from 616 kg/m³ at the base of the reservoir (layer 1) down to 355 kg/m³ at the top of the Utsira reservoir (Layer 9).

A final issue, to set the scene for understanding CO₂ volumes and masses, is to relate the quantities stored underground to the equivalent amounts emitted to atmosphere—which is, after all, the primary motivation for doing CO₂ storage. Sleipner, as a type-example CO₂ storage project, has stored approximately 1 million of tonnes per annum (Mtpa) of CO₂ and other more recent CCS projects such as the Boundary Dam and Quest projects in Canada also have a similar annual target. Note, the actual injection at Sleipner over the first 20 years varied from 0.6 to 1.0 Mtpa depending on the amount CO₂ available for storage. So how much CO₂ emission corresponds to 1 million tonnes? Table 2.2 summarises selected emissions data for various forms of transport. Of course, many efforts are underway to reduce CO₂ emissions from transport, so it depends where and when you make the measurement and which type of vehicle is assumed. However, in general we can conclude that 1 Mtpa stored in a CCS project is approximately equivalent to:

- Annual emissions from about 330,000 cars (assuming 200 g/km)
- 5 million passenger air kilometres
- 5 million tonnes of shipping for an Asia-EU transport distance
- One tenth of Norwegian road traffic emissions in 2014.

Table 2.2 Emissions data for various forms of transport

Emission type	CO_2 mass	Comment	Source
Air travel	113–257 g/km	CO_2 equivalent per passenger kilometre	Finland 2008 database; LIPASTO system (range is for long- to short-haul flights)
Vehicle emissions	200 g/km	Typical mid-size car in 2001	www.gov.uk (CO_2-and-vehicle-tax-tools) and US EPA average for 1975–2014
Vehicle emissions	118 g/km	EU average new car sold in 2016	ec.europa.eu/clima/policies/transport/
Vehicle emissions	95 g/km	EU 2021 target for average vehicle emissions	ec.europa.eu/clima/policies/transport/
Vehicle emissions	280 g/km	Truck	US EPA average figure for 1975–2014
Vehicle emissions	4.7 Mt/year	Typical annual vehicle CO_2 emissions	US EPA assumptions
Vehicle emissions	3 tonnes/year	Annual emissions for 15,000 km at 200 g/km	European/metric reference case
Vehicle emissions	10.2 Mt/year	Total 2014 Norwegian road traffic emissions	www.ssb.no/en/natur-og-miljo/statistikker/klimagassn
Maritime shipping (diesel)	10–15 g/tonne/km	Emissions of CO_2 per transport unit	World Shipping Council and https://www.eea.europa.eu/
Rail freight (diesel)	20–35 g/tonne/km	Emissions of CO_2 per transport unit	https://www.eea.europa.eu/

Industrial-scale CCS projects, like Sleipner, are thus clearly valuable and effective in significantly reducing a nation's emissions, further underpinning the motivation for CCS outlined in Chap. 1.

2.4.2 National and Regional Capacity Mapping Projects

Several national and continental studies have now been completed to map potential CO_2 storage formations and to estimate the storage capacity. A selection of the most mature studies includes:

- The EU GeoCapacity Project on European Capacity for Geological Storage of Carbon Dioxide (2008; http://www.geology.cz/geocapacity);
- The North American Carbon Storage Atlas covering USA, Canada and Mexico (2012; www.nacsap.org);
- The CO_2 atlas for the Norwegian Continental Shelf (2014; www.npd.no/en/Publications/Reports/Compiled-CO2-atlas/);
- Other national CO_2 storage databases including UK (www.co2stored.co.uk), Australia and Brasil.

These national, mainly government-sponsored, projects have set out to prepare nations for future large-scale CO_2 storage activities and in general they conclude that there is plenty of theoretical capacity available. For example, the North American estimate is that there is over 2400 billion metric tons (Gt) of theoretical storage capacity, and the Norwegian Continental Shelf has over 80 Gt potentially available (based on cut-off criteria).

However, there is also much debate about how realistic these preliminary estimates are. It is therefore important to understand the different types or classes of CO_2 storage capacity estimate. Bachu et al (2007) provide valuable reviews of the methods used in CO_2 storage capacity estimation.

There are several different types of estimate which are best summarized by the techno-economic resource–reserve pyramid (introduced by Bradshaw et al. 2007). Using the terms proposed by Bachu et al. (2007) (Fig. 2.17) we can differentiate:

- Theoretical capacity (the physical limit);
- Effective capacity (an estimate using cut-off criteria);
- Practical capacity (taking into account economic, technical and regulatory factors);
- Matched capacity (site-specific storage for specific CO_2 projects).

Various adaptations of this scheme have been used, such as to define capacities for the exploration, appraisal and development phases (used by the Norwegian CO_2 atlas). Also, several authorities are working on more formalized capacity definitions (including the UNFCCC, the SPE and ISO), so that some evolution of capacity definitions can be expected. For the sake of clarity here we will use the terms as defined by Bachu et al (2007) and focus on the methods involved.

Fig. 2.17 Techno-economic resource–reserve pyramid for discussing CO_2 storage estimates (from Bachu et al. 2007; © Elsevier, reproduced with permission)

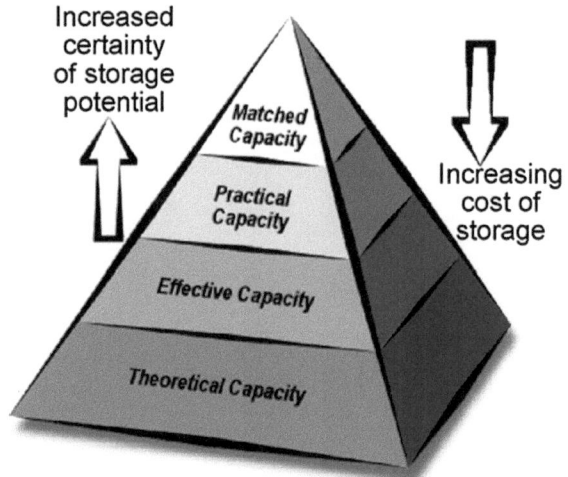

The resource pyramid is also dynamic, whereby as projects mature they move up the pyramid with increased certainty of storage potential (Fig. 2.17). Conversely, as the technology improves and as costs come down the matched or practical capacity fractions may be extended downwards. Ultimately, it is matched capacity that is the determining factor. CO_2 storage targets need to be quantified and matched with specific delivery of CO_2 from major source points (capture from power plants and industrial sources). Figure 2.18 shows an example assessment of CO_2 emissions, infrastructure and storage capacity for NW Europe, where the North Sea basin is the primary storage resource.

2.4.3 Methods for Storage Capacity Estimation

The *theoretical* CO_2 storage capacity, V_{CO_2}, for a structural or stratigraphic trap in a porous geological formation can be simply estimated in terms of the pore volume available for storage:

$$V_{CO_2} = V_{trap}\phi(1 - S_{wirr}) \qquad (2.6)$$

where V_{trap} is the volume of the trap, ϕ is the porosity and S_{wirr} is the irreducible water saturation.

It is also conventional for sandstone systems to discount the non-sand fractions using a net/gross ratio, and generally calculations in mass of CO_2 are preferred to volume (obtained by using the in situ density, ρ_{CO_2}). Theoretical CO_2 storage capacity could also include volumes below the structural closures, where injected CO_2 is

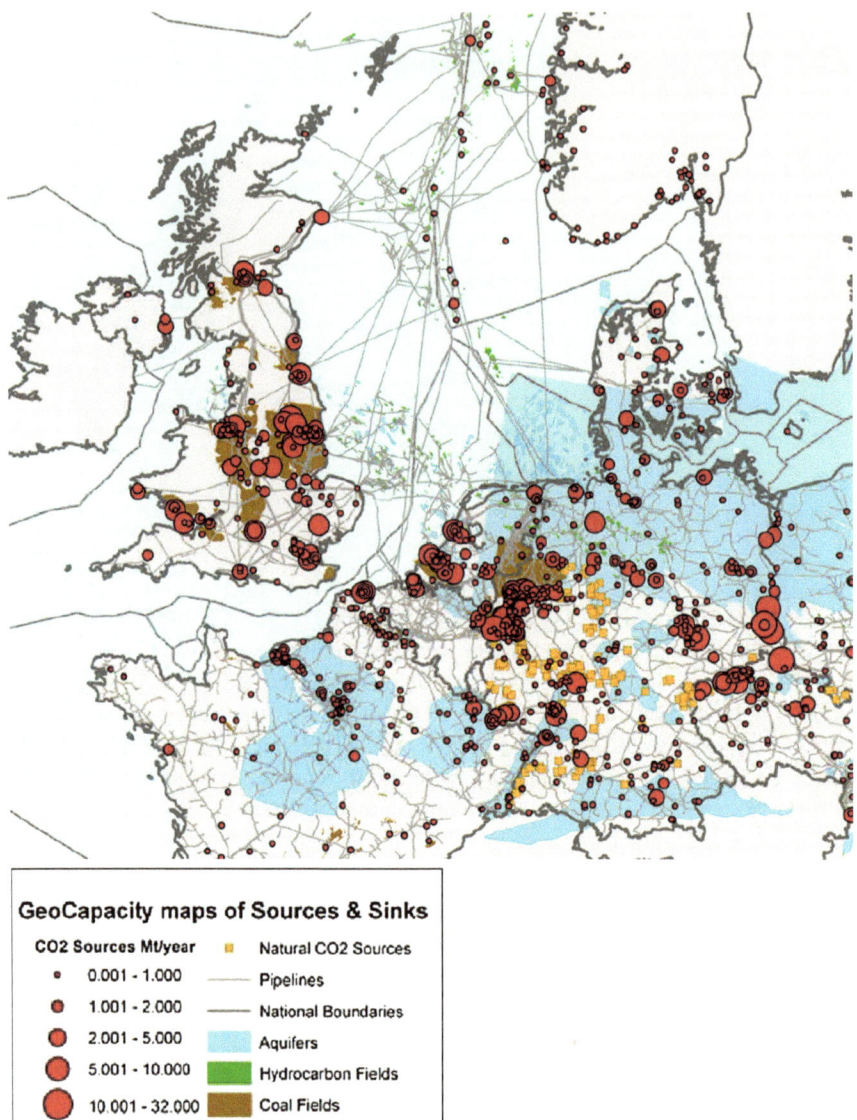

Fig. 2.18 Map of CO_2 emissions, infrastructure and storage capacity in NW Europe (downloaded from www.geocapacity.eu; EU Geocapacity final report, 2009)

trapped as a residual phase after migrating through a porous medium from some deeper injection point, and this residual-phase storage component can be estimated by:

$$V_{CO_2} = V_{swept}\phi S_{CO_2R} \tag{2.7}$$

where V_{swept} is the volume contacted by migrating CO_2 and S_{CO2R} is the residual CO_2-phase saturation.

Note that the irreducible water saturation, S_{wirr}, and the residual CO_2 saturation, S_{CO2R}, are highly dependent on the rock type and pore size, but typical values might be $S_{wirr} = 0.2–0.4$ and $S_{CO_2R} = 0.2–0.3$ (Fig. 2.10).

These theoretical estimates of the 'available pore space' for CO_2 storage do not include the effects of fluid dynamics, whereby the CO_2 will only fill a fraction of the available pore space. Therefore, a storage efficiency factor, ε, has been introduced to give a widely-used expression (Bachu 2015) for the *effective* storage capacity of the available pore volume, V_ϕ, within the saline aquifer:

$$M_{CO_2} = V_\phi \rho_{CO_2} \varepsilon \tag{2.8}$$

Thus, if the geometries and porosities of targeted geological units in the saline aquifer are known (using conventional subsurface mapping methods), then the storage capacity, M_{CO2}, can be estimated from factors that can be reasonably estimated from exploration and appraisal wells. Typically, practical decisions need to be taken on scaling the bulk rock volume, V_b, by the porosity, the net to gross ratio and the maximum CO_2 saturation at the pore-scale $(1 - S_{wirr})$, such that a more general version of Eq. 2.8 would be as follows:

$$M_{CO_2} = V_b \phi N/G \, \rho_{CO_2} \varepsilon \, (1 - S_{wirr}) \tag{2.9}$$

Different regions and practitioners may make different assumptions about what to include (Bachu, 2015), so it is important to be aware of the assumptions made. However, Eqs. 2.8 and 2.9 represent the essential concepts. The storage efficiency factor, ε, is the ratio of the actual volume of CO_2 stored in the aquifer volume to the theoretical (estimated) pore volume available (van der Meer 1995) and represents the cumulative effects of heterogeneity, fluid segregation and sweep efficiency. However, ε is difficult to estimate and is very site specific. General experience suggests that ε is in the range of 0.005–0.06 (i.e. less than 6% of the pore volume). Storage efficiency can also be estimated using reservoir simulation or analytical approaches, as discussed below.

For CO_2 storage in depleted hydrocarbon reservoirs, the hydrocarbon volume of the geological structure may be quite well quantified from historical production well data and seismic surveys. In these cases, it can be useful to estimate the storage capacity from the known hydrocarbon volume initially in place (HCIIP):

$$M_{CO_2} = HCIIP \, \rho_{CO_2} R_f (1 - F_{ig}) B_{HC} \tag{2.10}$$

where R_f = Recovery factor (the fraction of hydrocarbon produced), F_{ig} = Fraction of injected gas (in cases where gas injection has been used), B_{HC} = Hydrocarbon formation volume factor (the correction for gas or oil volumes measured at surface).

In most cases, it is depleted gas fields that are of primary interest for storage, but these gas fields typically contain some liquid hydrocarbons. Depleted oilfields are more likely to be targets for CO_2 storage associated with CO_2 EOR, where such simple volumetric calculations are not valid.

2.4.4 Understanding Storage Efficiency

Accurate estimates of CO_2 storage volumes and the corresponding storage efficiency factor, ε, are best done using dynamic flow simulation models and detailed 3D geological reservoir models. However, analytical modelling approaches are a useful and quick way to get an appreciation of the likely efficiency of CO_2 storage.

In general, we know that a less dense fluid (gas or CO_2) will rise due to buoyancy and migrate up-dip to occupy only a small fraction of the aquifer layer pore volume (as sketched in Fig. 2.8). This effect can be captured analytically using the characteristic geometry of a CO_2 plume (or any two-phase, immiscible fluid system). Nordbotten et al. (2005) applied these concepts for CO_2 storage in deep saline aquifers using approaches previously applied to hydrocarbon-water systems (e.g. Rapoport 1955; Shook et al. 1992; Ringrose et al. 1993). Here we will summarize this approach as an introduction to analytical methods which are now widely applied to this problem (e.g. Okwen et al. 2010).

For a vertical well injecting at a rate Q_{well} into a horizontal saline aquifer unit, thickness B, the CO_2 plume will expand with a 'curved inverted cone' geometry with a radius, r (Fig. 2.19a). Nordbotten et al. (2005) developed an analytical solution to this problem characterized by the geometry shown in Fig. 2.19b.

The actual shape of the plume will depend on many factors, especially the CO_2 density and the flow rate. A larger density difference ($\rho_{brine} - \rho_{CO_2}$) will give more vertical migration and lateral spreading (a larger r_{max}), while a higher fluid flow rate will promote more viscous flow in the near well bore region (a larger r_{min}). In terms of fluid dynamics, the shape of the curve is controlled by the ratio of viscous to gravity forces; Nordbotten et al. (2005) and Nordbotten and Celia (2006) showed how the shape of the curve varies as a function of the gravity/viscous ratio and other dimensionless ratios.

This analytical model can be used to estimate the storage efficiency. Here it is useful to define a CO_2 storage capacity coefficient, C_c, for an expanding cylinder containing the plume (Fig. 2.20):

$$C_c = V_{injected}/V_{PV} \tag{2.11}$$

where V_{PV} is the total pore volume of the cylinder.

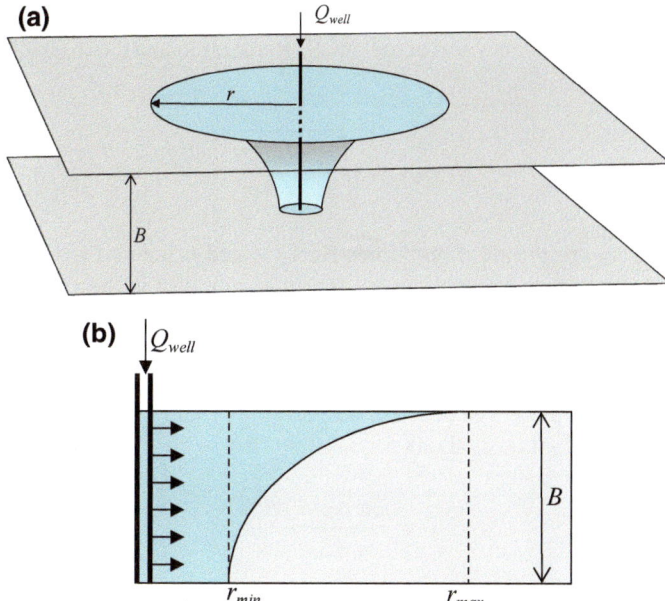

Fig. 2.19 **a** Ideal geometry of a CO_2 plume expanding as buoyant fluid in a deep saline aquifer. **b** Terms used to define analytical plume geometry (based on Nordbotten et al. 2005)

Fig. 2.20 Factors defining the storage capacity coefficient, C_c, for an expanding cylinder containing a CO_2 plume

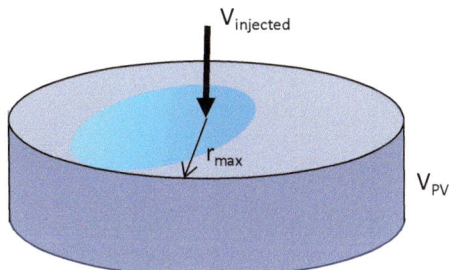

Note that at the end of injection C_c is equivalent to the final storage efficiency, ε. Of course, for real cases the plume may be any shape, but for the analytical solution the plume is assumed to be circular.

Since we know the pore volume of a cylinder, the storage capacity coefficient, C_c is given by:

$$C_c = \frac{V_{injected}}{V_{PV}} = \frac{Q_{well}t}{\phi B \pi (r_{max})^2} \qquad (2.12)$$

For a real storage site, r_{max} could be determined from monitoring data (e.g. first breakthrough to a monitoring well or by using time-lapse seismic images). For a predictive case, we might like to estimate this analytically. When the flow is viscous-dominated, and the buoyancy forces are small, Nordbotten and Celia (2006) showed that r_{max} is given by:

$$r_{max} = \sqrt{\frac{\lambda_c}{\lambda_b} \frac{Q_{well}t}{\pi B \phi}} \qquad (2.13)$$

where λ_c and λ_b are the fluid mobilities for CO_2 and brine and t is the injection time interval. Note that for each phase, the fluid mobility is the ratio of relative permeability to viscosity, $\lambda_i = k_i/\mu_i$. Fluid mobility is also usefully summarized by the mobility ratio $\lambda_r = \lambda_c/\lambda_b$.

This solution is for the case of viscous-dominated flow in a horizontal continuous aquifer, where the Gravity/Viscous ratio is small ($\Gamma < 1$), and where the gravity factor, Γ, is given by (Nordbotten et al. 2005):

$$\Gamma = \frac{2\pi \Delta\rho k \lambda_b B^2}{Q_{well}} \qquad (2.14)$$

where k is the permeability $\Delta\rho$ is the fluid density difference.

It is clear that estimation of C_c using this method gives a result which is very dependent on the fluid properties. For example, for the case of injecting CO_2 at the depth of around 1 km into a 100 m aquifer, the analytical value for C_c lies is around 0.25 (assuming a mobility ratio of $\lambda_r = 4$ and $\Delta\rho = 300$ kg/m^3). However, the value is very dependent on the mobility ratio, as shown in Fig. 2.21.

By substituting Eq. 2.12 into Eq. 2.13, we can infer that, for a viscous-dominated case:

$$C_c = \frac{1}{\lambda_r} = \frac{\lambda_b}{\lambda_c} = \frac{k_{rb}}{k_{rc}} \frac{\mu_c}{\mu_b} \qquad (2.15)$$

where, k_{rb} and k_{rc} are the relative permeabilities for brine and CO_2.

Thus, C_c is in fact the inverse of the mobility ratio, λ_r, for this limiting case. Note that since k_{rb} and k_{rc} vary as a function of saturation, the mobility ratio is also a variable function. For scoping calculations it is conventional to use the endpoint relative permeability values (ref. Figure 2.10).

However, neglecting the effect of gravity is rather misleading, since at CO_2 storage conditions the effects of gravity are significant—CO_2 is significantly lighter and less viscous than water. Okwen et al. (2010) extended this analytical approach to evaluate storage efficiency as a function of the gravity factor, Γ (Fig. 2.22). Their analysis shows how as soon gravity effects are included the storage efficiency reduces considerably. Typical values for CO_2 storage sites give $10 < \Gamma < 50$, implying values for ε generally less than 0.06. This then is the underlying theoretical reason why typical estimates for storage efficiency are $0.01 < \varepsilon < 0.06$. After 20 years injection at the Sleipner site, we can measure the actual storage efficiency based on time-lapse

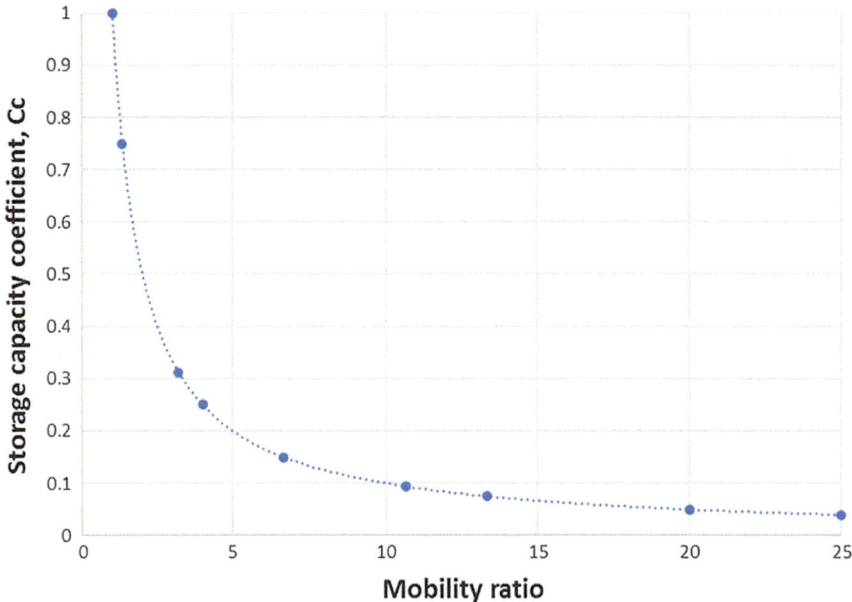

Fig. 2.21 Storage capacity coefficient, C_c, as a function of mobility ratio, λ_r

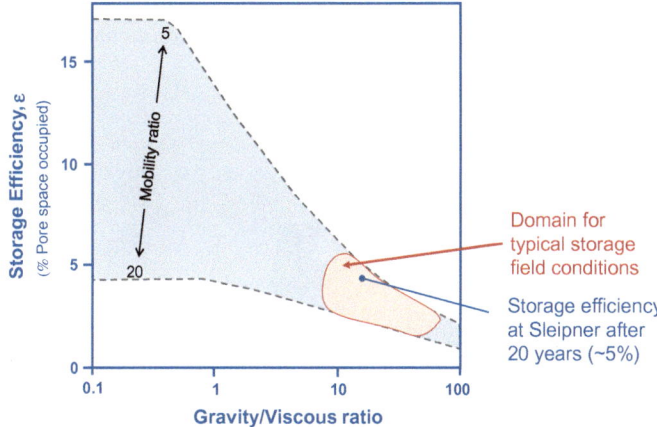

Fig. 2.22 Storage efficiency, ε, as a function of mobility ratio, λ_r, and gravity factor, Γ (modified from Okwen et al. 2010)

seismic data, and the estimates give $\varepsilon \sim 5\%$, confirming the theoretical framework is correct (Ringrose 2018). This is discussed further below.

2.4.5   Illustration of Capacity Estimates for the Sleipner Case

The Sleipner case provides a very informative example of the factors which control CO_2 storage capacity estimates in practice. CO_2 storage at Sleipner has been in operation since 1996 (Baklid et al. 1996; Eiken et al. 2011) and provides a unique insight into the dynamical behaviour of a CO_2 plume in the subsurface, having been monitored by regular time-lapse seismic surveys over a period of 20 years (Arts et al. 2004; Furre et al. 2015, 2017).

Figures 2.23 and 2.24 summarise the time-lapse seismic observations which can be used to determine the plume growth geometry. The uppermost layer (layer 9) has the best seismic imaging quality, as it avoids the complexities of seismic wave interference and time-delays which affect the lower layers. Overall the CO_2 plume shows an 'inverted cone' geometry with a more complex pattern of plume expansion in different layers within the 200 m think sandstone. Singh et al. (2010) presented reservoir models and supporting data for modelling the CO_2 plume in Layer 9. Using these data, together with the seismic amplitude maps, we can then estimate CO_2 efficiency parameters for a real case. Estimates for the plume extent and mass rate are given in Table 2.3. These are based on seismic observations and neglect possible portions of the plume not detected on seismic.

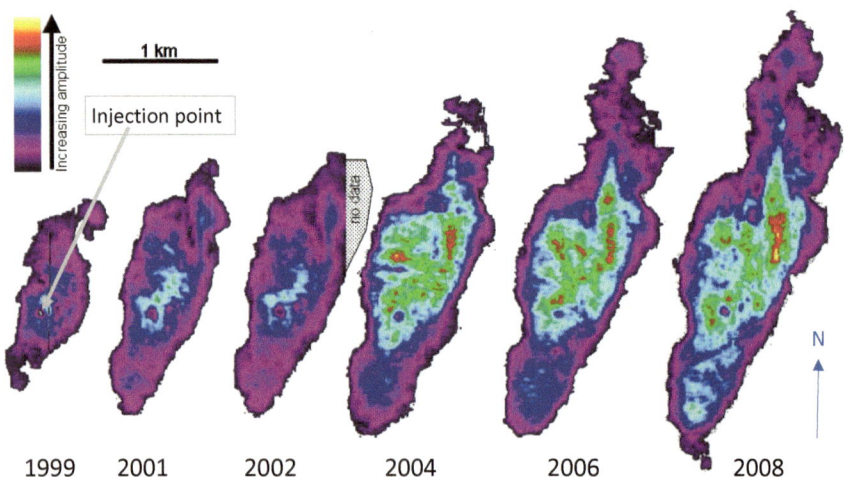

Fig. 2.23 Time-lapse seismic difference reflection amplitude maps at Sleipner (cumulative for all layers) showing expansion of the plume in all directions with largest extension to the north and intensified reflections in the central part of the plume (from Eiken et al. 2011, © Elsevier, reproduced with permission)

Fig. 2.24 Seismic sections (N-S) at Sleipner showing pre-injection conditions (1994), enhanced reflection amplitudes due to CO_2 invasion into multiple layers by 2008 and time-lapse difference reflection amplitudes (2008–1994). (Image courtesy of Equinor)

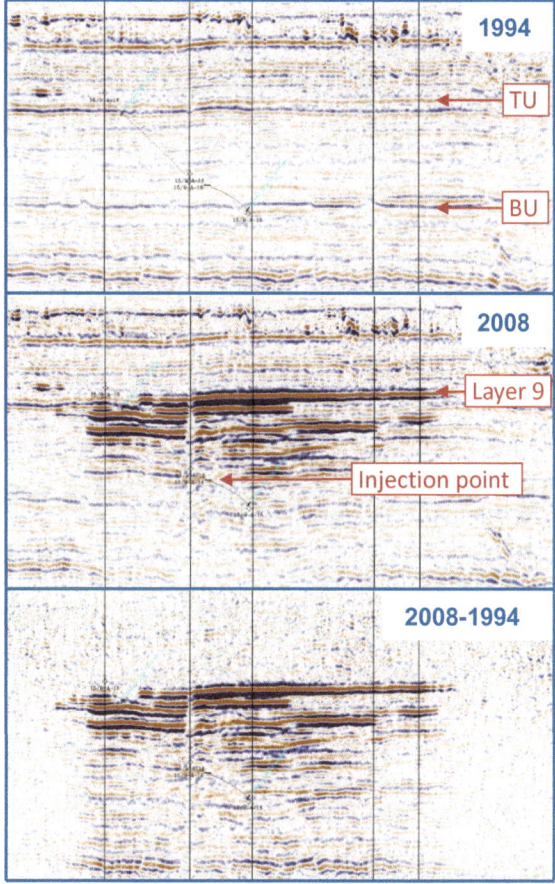

Table 2.3 Plume growth observations for Sleipner Layer 9. R_{max} is the distance from the injection point to the northernmost extension of the plume (from seismic data) and the mass injected into Layer 9 is based on a fraction of the total injection estimate from amplitude maps (modified from Singh et al. 2010)

Date	R_{max} (m)	Mass into Layer 9 (Mt)
1996	0	0.00
1999	346	0.02
2001	1022	0.10
2002	1290	0.16
2004	2091	0.33
2006	2569	0.66
2008	2786	1.19
2010	2942	1.83

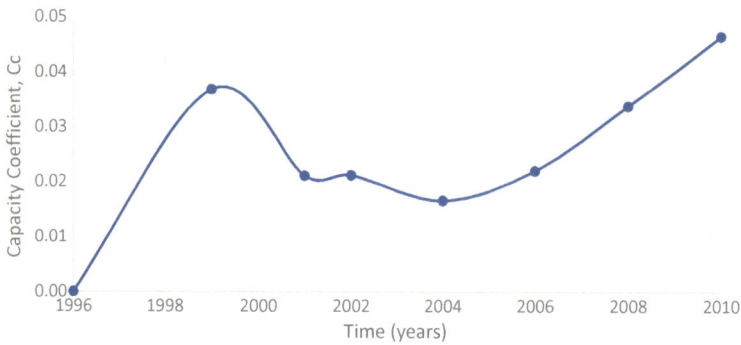

Fig. 2.25 Estimates of the storage capacity coefficient, C_c, for layer 9

Figure 2.25 shows the storage capacity coefficient, C_c (Eq. 2.12) for layer 9 up to year 2010. Here we have assumed a reservoir thickness of 11.3 m, a porosity of 0.36 and a CO_2 density of 355 kg/m^3 (based on estimates from Singh et al. 2010 and Cavanagh et al. 2015). The value for C_c is observed to rise to 0.04 then decrease before rising again to nearly 0.05. This variation is mainly caused by the top surface topography whereby one structural closure is filled by CO_2 before the plume then expands into a second structural closure to the north. Overall, the storage capacity coefficient for Layer 9 seems to be close to 0.05.

We can also estimate the overall storage efficiency ε at the Sleipner site based on historical monitoring observations. Using the 2013 seismic monitoring dataset as a reference (Furre et al. 2017), and assuming Eq. (2.8) we can estimate a site storage efficiency at Sleipner of $\varepsilon \sim 5.2\%$ (in 2013), using the following assumptions:

- The total quantity of CO_2 injected by 2013 was 14.7 Mt;
- The storage unit volume is defined by an area of 4 by 1.5 km covering the structural closures and plume area (revealed from seismic amplitude maps from the 2013 survey).
- This area corresponds to a total available pore volume of 4.17×10^8 m^3.
- The CO_2 density varies as a function of depth (Cavanagh et al. 2015); however based on gravity data, we can estimate an average density of 675 kg/m^3 (Alnes et al. 2011).

This storage efficiency is achieved thanks to the trapping of the CO_2 beneath several intermediate shale layers and due to the existence of local structural closures that contain the CO_2 plume. Furre et al (2019) show that there is evidence for multiple vertical feeder points within the CO_2 plume at Sleipner, and so the true multiple-layer multiple feeder point filling dynamics is undoubtedly complex (ref. Figure 2.24). Despite this internal complexity, the gradual filling of the sandstone layers and containment within structural closures gives a fairly linear function for ε (Fig. 2.26) reaching around 0.52 by year 2013.

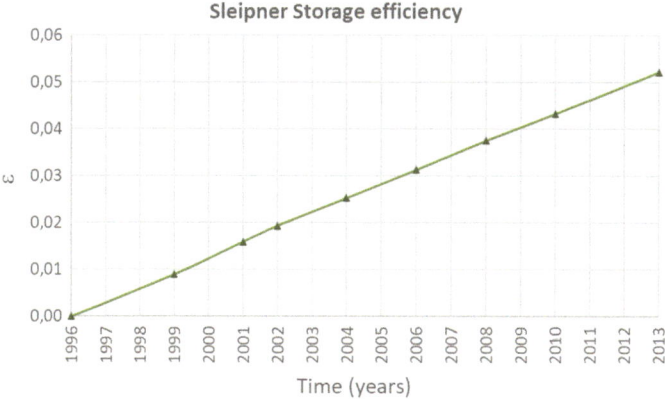

Fig. 2.26 Estimates of the storage efficiency factor, ε, for the whole of the Sleipner Utsira storage unit

2.5 CO₂ Storage Flow Dynamics

2.5.1 Dimensionless Ratios for CO₂ Storage

Before we start modelling or predicting CO_2 flow dynamics in the subsurface we need to understand the physical processes operating. Understanding the fluid forces is rather important to understanding CO_2 storage, so a summary of the concept of *dimensionless ratios* is important. Injecting CO_2 into a brine-filled permeable rock formation is part of a class of two-phase flow problems which have been extensively studied. A two-phase CO_2-brine system is analogous to two-phase oil-water system or a gas-water system. In each of these systems the fluid pair is generally immiscible (they do not mix), while other fluid pairs such as gas-oil are miscible. CO_2-oil systems, of importance to CO_2 EOR schemes, can be both miscible and immiscible, depending on depth, pressure and temperature. Real fluid systems may also be three-phase (e.g. CO_2-brine-oil), so things can get very complicated. However, it is informative here to just focus on the nature of two-phase immiscible flow problems as generally applicable to a CO_2-brine system.

The underlying concept is that by defining ratios of forces acting on the fluids, we can derive dimensionless numbers that characterize the flow dynamics (e.g. Rapoport 1955). There are many ways of defining such systems, and the mathematics can become quite extensive. Here, by restricting our discussion to the most important ratios, we introduce the main concepts and identify the key factors which control a two-phase immiscible flow system (Shook et al. 1992; Ringrose et al. 1993). Firstly, we define fluid force ratios, by determining the ratio of pressure gradients causes by viscous, capillary and gravity forces.

By Darcy's law, the viscous force in the direction x is proportional to viscosity, μ, permeability, k_x and fluid velocity u_x, as follows:

$$\frac{dP}{dx} \propto \frac{\mu}{k_x} u_x \tag{2.16}$$

The law can be applied to each phase and the pressure gradient may be in any direction (generally requiring a full vector field expansion of Eq. 2.16 and permeability as a tensor). The capillary force is defined by capillary pressure as a function of water saturation (dP_c/dS_w), and the gravity force is defined by the fluid density difference and the acceleration due to gravity, g, in the vertical direction z, and is given by $\Delta \rho g \Delta z$.

Using ratios of pressure gradients, the characteristic dimensionless ratios can be most simply defined for a 2D cross section with height, Δz, and length, Δx, as follows:

$$\text{Viscous/capillary ratio} \frac{u_x \Delta x \mu_{nw}}{k_x (dP_c/dS_w)} \tag{2.17}$$

$$\text{Gravity/capillary ratio} \frac{\Delta \rho g \Delta z}{(dP_c/dS)} \tag{2.18}$$

In addition, we need to define a geometrical ratio describing the anisotropy of the system and the viscosity ratio:

$$\text{System shape group } \frac{k_x}{k_z} \left(\frac{\Delta z}{\Delta x} \right)^2 \tag{2.19}$$

$$\text{Viscosity ratio } \frac{\mu_{nw}}{\mu_w} \tag{2.20}$$

where μ_{nw} and μ_w are the viscosities of the non-wetting and wetting phases (in our case, CO$_2$ and water).

These ratios can be used to understand how a CO$_2$-brine flow system behaves under certain conditions; for example, whether flow conditions are dominated by viscous, gravity or capillary forces or what happens as the system becomes highly anisotropic. For a given set of reservoir conditions at the CO$_2$ storage site, we can use these ratios to quickly determine which factors are likely to be most critical. As argued above, most CO$_2$ storage processes are expected to be gravity dominated.

However, because the forces vary as a function of time, location and direction (Stephen et al. 2001), it is usually necessary to define more precise dimensionless ratios for specific situations. For example, Zhou et al. (1997) proposed a modified set of dimensionless ratios for the case of horizontal flow in a 2D vertical section, to give the following dimensionless numbers for the viscous/capillary ratio, N_{vc}, and the gravity/viscous ratio, N_{gv}:

$$N_{VC} = \frac{u_x \Delta z^2 \mu_{nw}}{k_z \Delta x (dP_c/dS_w)} \tag{2.21}$$

$$N_{GV} = \frac{\Delta \rho g \Delta x k_v}{u_x \Delta z \mu_{nw}} \tag{2.22}$$

Notice that Eq. 2.21 and Eq. 2.17 are both dimensionless ratios, but the length terms are different because of the way in which the shape group is handled. Another useful set of assumptions for a CO$_2$-brine flow system is the condition of vertical equilibrium, where the two fluids are assumed vertically segregated due to gravity. Yortsos (1995) showed that vertically segregated flow will tend to occur when, $N_{GV} \ll 1$ (Eq. 2.22).

A widely-used and more fundamental viscous/capillary ratio is the Capillary number, C_a:

$$C_a = \mu u / \gamma \tag{2.23}$$

where γ is the interfacial tension.

Another important gravity/capillary ratio is the Bond number (or Eötvös number), given by:

$$B_o = \Delta \rho g L^2 / \gamma \tag{2.24}$$

where L is a characteristic length.

In the context of CO$_2$ storage, C_a and B_o, are most typically used at the pore-scale when assessing the physical behaviour of a droplet of CO$_2$, while N_{VC} and N_{GV} are more commonly used when assessing large-scale macroscopic flow problems.

We can use these dimensionless ratios to determine what kind of fluid dynamics we can expect to operate in the CO$_2$ storage reservoir. Close to the well, where pressure gradients are high, we can expect viscous-dominated conditions, while in the reservoir region away from the well we find that gravity-dominated conditions will occur.

For the viscous/capillary ratio, N_{vc}, quantifying the capillary force can be quite challenging. For homogeneous porous media, capillary forces only operate at very small scales (at the pore-scale and up to around 0.2 m) and have little impact at larger scales. However, because reservoir rock formations are heterogenous (especially due to lamination and bedding) the effects of capillary forces can be quite significant at larger scales (Ringrose et al. 1993; Krevor et al. 2011). This effect is referred to as heterogeneity trapping, whereby small-scale heterogeneities (e.g. layering at the scale of 0.01–0.1 m) cause retention of the non-wetting phase due to capillary forces. Huang et al. (1995, 1996) demonstrated and quantified this effect in the laboratory for water-oil systems in laminated sandstones, where between 30 and 55% of the oil volume becomes trapped in isolated high-permeability laminae. Similar effects have been demonstrated for CO$_2$-brine systems (Reynolds and Krevor 2015; Trevisan et al. 2015), and models of CO$_2$ storage systems that account for heterogeneity trapping demonstrate that a significant amount of CO$_2$ storage is likely to be in the form of residual CO$_2$ saturation (Krevor et al. 2015; Meckel et al. 2015).

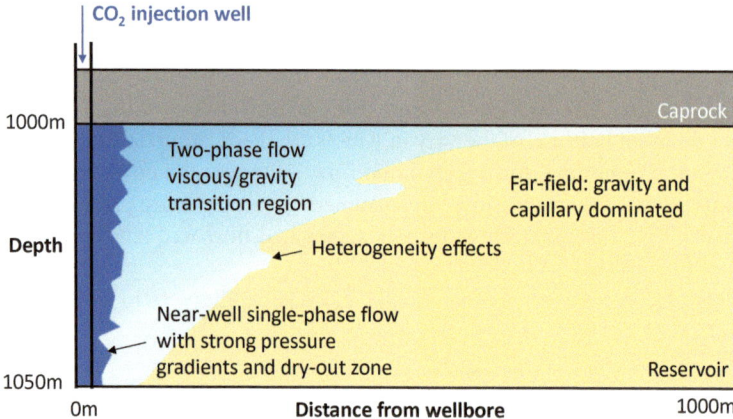

Fig. 2.27 Sketch of flow processes and flow regimes for CO_2 injection into an idealized storage unit (modified from Oldenburg et al. 2016)

Although the relative role of viscous, capillary and gravity forces in a CO_2 storage unit will be site specific, some general principles can be established (Fig. 2.27). High pressure gradients are expected in the near-well region, where single-phase flow and dry-out effects are expected. The dry-out effect in the near-wellbore region is caused by CO_2 changing the solubility of salts in the in situ brine and will be discussed later in the section on CO_2 storage project design. In the zone c.100 m around the well we can expect a two-phase flow region with a gradual transition from viscous- to capillary- dominated flow, and in the far-field region >100 m from the well we expect gravity and capillary effects to dominate. Figure 2.27 illustrates these effects, consistent with the general concept captured by the analytical model (Fig. 2.19) but adding the complexities caused by varying pressure gradients and changing dynamic flow regimes. In the following section we will discuss approaches for modelling these effects (now that we have an idea of what we are trying to model).

2.5.2   *Approaches for Modelling CO_2 Storage*

As soon as we move beyond the conceptual aspects of CO_2 storage to start looking at specific storage sites, we will need to use 3D numerical simulation models. This is because the geological architecture and the flow processes are generally so complex that some form of detailed forecasting is needed. Geological reservoir modelling is a major topic itself and is covered elsewhere (e.g. Ringrose and Bentley 2016). Here we will summarize the main methods used, focusing on critical factors for achieving fit-for-purpose models of CO_2 storage sites.

Essentially, we need two things:

- A 3D geological model (the static model)

• A fluid flow simulation model (the dynamic model).

The geological model could be simply a 'container' describing the top and bottom stratigraphic surfaces along with the main bounding faults and some 'best-estimates' of porosity and permeability, in which case a limited amount of seismic and well data could be sufficient to define a relatively simple geological model. More likely, a more complex, multiple-layer rock model with information on litho-facies distributions, structural heterogeneities and petrophysical properties will be needed. Ringrose and Bentley (2016) describe the workflows and processes involved in model building, stressing the importance understanding the balance between stochastic and deterministic aspects of modelling and the multi-scale nature of rock systems. Figures 2.28 and 2.29 show examples of static rock property models and dynamic flow simulation models from the CO$_2$ storage projects at In Salah, Algeria, and Snøhvit, Norway.

In practice, several alternative geological models may be needed, and as the factors controlling the flow problem become better understood, models will need to be updated and refined. Historically, geological reservoir modelling has been very constrained by the time and cost involved in building static grid models, with a tendency to 'lock' the geological model concept into one poorly-defined (and often misleading) grid. This should be avoided, and it is always advisable to maintain alternative models and concepts to ensure the range of possible outcomes is appreciated. In future, with rapid advances in modelling technology, we can expect the modelling process to be much more flexible and grid-independent, allowing modellers to quickly pose

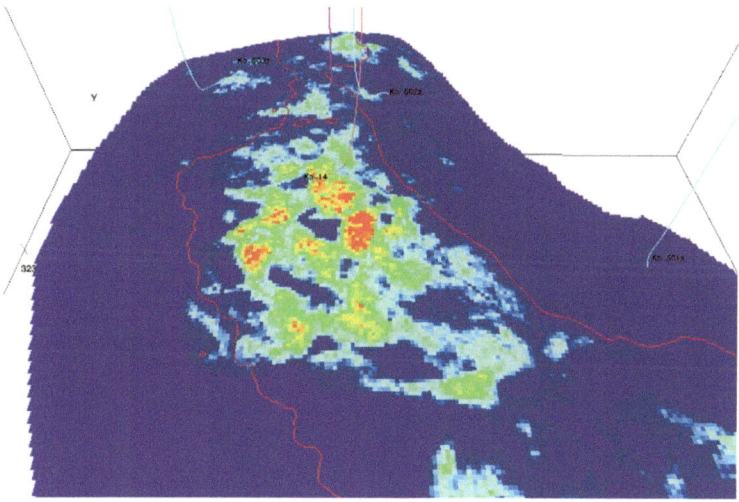

Fig. 2.28 Example geological model of a CO$_2$ storage site, showing a porosity map estimated from seismic amplitude data draped on the base reservoir surface (hot colours indicate porosity up to 20%); In Salah CO$_2$ project in Algeria (Ringrose et al. 2011). Red outline is the initial gas-water contact for the gas field and blue wells are CO$_2$ injection wells into the down-dip saline aquifer. Model cube is c. 15 km wide

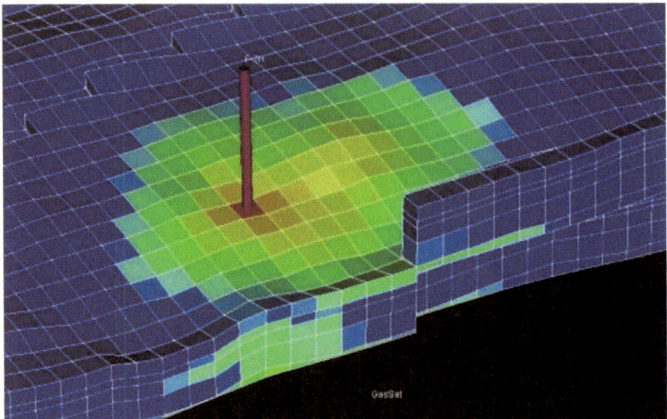

Fig. 2.29 Example flow simulation model showing CO$_2$ distributions (green and yellow shading) in a coarse-grid scoping model of the Snøhvit CO$_2$ injection site (Hansen et al. 2013). Here, the 80 m-thick Tubåen reservoir interval is represented by five simulation grid layers, with bounding faults represented by transmissibility multipliers to study possible cross-fault flow

different geological concepts and scenarios and test them in terms of the implications for fluid flow (e.g. Bentley and Ringrose 2017).

Assuming we have a defined 3D geological model we want to work with, the next step is to appreciate the type of flow simulation model we wish to use. Again, numerical flow simulation is an extensive topic with many components involved, from flow physics to numerical methods. Here, we will summarize the main approaches used for CO$_2$ storage simulation with the aim of guiding the decision process and model choice. We need to know 'what kind of flow simulation method we might wish to use?'

We have already introduced analytical modelling approaches above (Sect. 2.4.4) and the concept of dimensionless ratios (Sect. 2.4.5) as a means for identifying which fluid forces are likely to be important. Since CO$_2$ storage modelling is essentially a two-phase flow problem, we need to know what type of two-phase flow simulation method we might need to use.

Firstly, we can summarize the flow problem we are trying to solve. For a two-phase immiscible flow problem, the two-phase Darcy flow equations are used:

$$u_w = \frac{\bar{k}k_{rw}}{\mu_w}(\nabla P_w + \rho_w g \nabla z) \qquad (2.25)$$

$$u_n = \frac{\bar{k}k_{rn}}{\mu_n}(\nabla P_n + \rho_n g \nabla z) \qquad (2.26)$$

where

u_w and u_n are the velocities of the wetting and non-wetting phases (water and CO$_2$)

$\bar{k}$ is the absolute permeability tensor,
k$_{rw}$ and k$_{rn}$ are the relative permeabilities,
μ_w and μ_n are the viscosities,
∇P is the gradient of pressure,
ρ_w and ρ_w are the fluid densities,
g is the acceleration due to gravity,
∇z is the gradient of the vertical co-ordinate.

A third equation is needed to solve the two-phase Darcy flow problem and that is the relationship between the phase pressures, known as the capillary pressure equation:

$$P_c = (P_n - P_w) = \frac{2\gamma \cos \theta}{r_{eff}} \qquad (2.27)$$

where σ is the interfacial tension, θ is the fluid contact angle and r$_{eff}$ is the effective pore radius.

Because the capillary pressure is very strongly dependent on the phase saturation, it is most commonly represented by the gradient with respect to saturation (dP$_c$/dS) as used in Eqs. 2.17 and 2.18. Most commonly, P$_c$ is assumed to be known from core measurements or empirical functions, and therefore the main numerical problem is to solve the two-phase Darcy flow equations. Another essential constraint in flow modelling is conservation of mass—the sum of all fluids flowing into and out of a model domain (e.g. a grid cell) must be zero. Fuller treatments of the theory behind the two-phase Darcy flow problem applied to CO$_2$ storage are given by Nordbotten and Celia (2011) and Niemi et al. (2017).

Here we will review the main methods for solving this two-phase flow problem. There are three main types of model method that are of most interest:

1. Two-phase finite-difference (control volume) methods
2. Two-phase finite-element methods
3. Invasion Percolation (IP) methods.

Figure 2.30 illustrates the main differences between these approaches. The finite-difference (FD) method is the dominant approach used in CO$_2$ storage flow simulation (as well as in the hydrology and petroleum engineering disciplines). This is because the FD method is generally the most efficient (computationally) and also robust in terms of accuracy and ease of obtaining a solution. FD methods use cell-centre nodes and then apply a set of numerical solutions to calculate multiphase-phase flow fluxes between cell centres. Because the two-phase flow problem is relatively easily solved using linear-matrix computational methods, FD methods applied to regular 3D Cartesian grids are generally efficient and versatile. The FD method is also well adapted to solving flow problems for heterogenous rock media (e.g. Durlofsky 1991; Pickup et al. 1994; Pickup and Sorbic 1996). The most widely used numerical recipe for doing this is the IMPES solution: IMplicit Pressure, Explicit Saturation. In this method, the pressures are calculated implicitly from the saturations which are

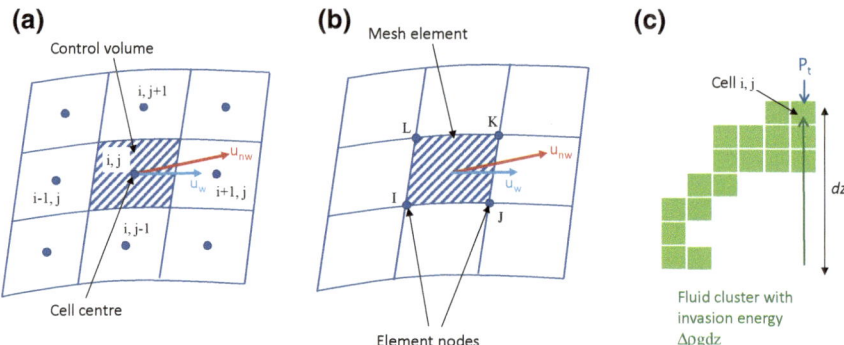

Fig. 2.30 Sketch illustrating different flow simulation methods using in modelling CO_2 storage: **a** finite difference method, **b** finite element method, **c** invasion percolation method

explicitly defined from the previous time step. FD methods require the grid to have a Cartesian geometry with only small departures from an orthogonal grid. Thus, a fairly regular, near-orthogonal grid is needed for FD flow simulation. The most common grid format is called 'corner point' where each 3D cell volume has 6 faces and 8 nodes (i.e. topologically equivalent to a cube).

There are, however, some limitations to the FD approach which lead us towards the more computationally demanding yet more flexible set of finite-element (FE) approaches. FE approaches allow higher flexibility on the grid geometry, which can allow cells to deform with time. The use of unstructured (or irregular) grids, such as triangles or tetrahedra, also enables more complex flow terms to be calculated (including geomechanical deformation, geochemical reactions and geothermal effects).

A finite element model involves grid meshes defined by nodes at each corner of the cell (Fig. 2.30b), such that complex geometries for cell deformation as a function of time can be more accurately represented. Typical practice in modelling CO_2 storage, is that FE methods are applied when solving specific geomechanical or geochemical problems, but not used so often for general flow simulation.

The third main approach for modelling CO_2 storage is Invasion Percolation (IP), where the Darcy equations are neglected in favour of a capillary/gravity dominated flow system using percolating clusters in a medium controlled by capillary threshold pressures. The IP concept, introduced by Wilkinson and Willemsen (1983), has been widely applied to multiphase flow problems at the pore scale (e.g. Bakke and Øren 1997; Blunt 2001), for modelling basin-scale oil migration (Carruthers and Ringrose 1998) and for modelling CO_2 storage (Cavanagh and Ringrose 2011; Cavanagh et al. 2015). When using the IP method to represent a migrating immiscible buoyant fluid invading a water-wet porous medium, the condition is used whereby migration of the non-wetting phase cluster occurs if:

$$\Delta \rho g dz > P_t \qquad (2.28)$$

where $\Delta\rho$ is the fluid density difference for the cluster with vertical dimension dz, and P_t is the threshold capillary pressure for largest available pore which can be invaded by the cluster.

If this condition is not satisfied, the cluster is immobile. The IP method thus uses gravity/capillary ratio (Eq. 2.18) or Bond number (Eq. 2.24) to control the flow behaviour, assuming viscous forces are negligible. Cavanagh and Haszeldine (2014) and Oldenburg et al. (2016) discuss the conditions where the IP method could be considered valid for representing CO$_2$ storage, essentially when the capillary number is very low. Generally, this will occur in the far field region, away from the influence of pressure gradients near injection wells (Fig. 2.27). The IP method is particularly well suited to modelling long-term CO$_2$ migration pathways where flow conditions will be gravity-dominated.

The above methods all relate to models of continuous permeable media. When it comes to handling faults and fractures, a further set of model variants emerges. Briefly, the options for modelling faulted and fractured reservoir systems are as follows:

- Relatively simple and continuous fault surfaces can be added as planes between grid cells in the FD approach (but if very complex fault networks are to be represented this can be very challenging);
- More complex fault or fracture systems can be modelled using FE approaches which allow much more flexibility on the grid design;
- Rock systems with many fractures can be treated with a dedicated approach know as a Discrete Fracture Network (DFN) model, where each fracture and its connection to other fractures is represented in a network of conducting planes within a background field (which is usually assumed to be impermeable).

Handling of faults and fractures in reservoir models is discussed further in Ringrose and Bentley (2016; Chap. 6).

2.5.3 Example CO$_2$ Storage Modelling Workflow

It is helpful to illustrate the steps in modelling a CO$_2$ storage site using a real example. Each storage site will have its own unique character and model requirements, so no general modelling process can be recommended. However, the modelling steps and decision-making process can be generalised to some degree, using the foundation that reservoir modelling should be a decision-led process (Bentley 2016; Bentley and Ringrose 2017). Here, we will use specific examples from the In Salah CO$_2$ storage demonstration project to illustrate how a range of models were used to address specific decisions related to managing a CO$_2$ storage project. Table 2.4 summarizes the links between models and decisions made for this case. The In Salah CCS project in central Algeria was part of the In Salah Joint Venture, a multi-field gas development project which included a CO$_2$ capture and storage project. The CO$_2$ storage demonstration

Table 2.4 Examples of decision-led flow modelling tasks for the In Salah case study

Question addressed	Type of modelling performed	Outcome of model	Decision
Which faults and fractures could affect CO_2 storage?	Structural modelling based on seismic data and using stress field analysis	Faults aligned with present-day stress field model likely to have associated fracture flow	Early CO_2 breakthrough to monitoring well is due to fault-related fractures
How do fractures control the early CO_2 breakthrough observed at well KB-5?	Reservoir flow simulation (FD) using modified permeabilities, informed by DFN models	Fracture permeability is much higher than matrix permeability and is dynamic (stress sensitive)	Modify injection plan and flow rates
How do fractures control flow throughout the storage unit?	Fracture (DFN) modelling of the entire reservoir	Improved estimates of fracture and matrix permeability tensors	Observed ellipsoidal uplift pattern is consistent with anisotropic (stress-controlled) flow model

project involved injection of captured CO_2 into the saline aquifer surrounding gas-bearing units in the 1.9 km deep Carboniferous sandstone unit (Fig. 2.31) at the Krechba field. The project injected 3.8 Mt of CO_2 in the period between 2004 and 2011 (Ringrose et al. 2013).

The geological setting of this storage site is important to appreciate before we discuss the modelling work. The Carboniferous storage unit at Krechba comprises a gentle anticline formed during compressive tectonic phases in the late Carboniferous Period (c. 300 Million years ago). A NE-SW compressive stress system deformed this Palaeozoic sedimentary basin (the Ahnet basin) into a series of folds (e.g. Fig. 2.32). Continued compression led to some of these folds being breached by strike-slip faults. The Krechba structure remained relatively un-faulted and the 20 m thick storage unit (C10.2) was never fully offset by faults. All the faults are thus subtle and close to the limits of seismic resolution. Substantial erosion, with up to 2 km of uplift, mainly represented by the Hercynian unconformity, led to stress relaxation and formation of joints. The present-day stress regime is strike-slip with a NW-SE maximum horizontal stress orientation. Understanding this rather complex structural history is an important backdrop to the work, summarized below, on inferring the nature of faults and fractures and their impact on CO_2 injection performance.

Figure 2.32 illustrates the work done to address the question 'Which faults and fractures could affect CO_2 storage?' The essential first step is mapping faults from 3D seismic—a set of detailed workflows, summarized by Ringrose et al. (2011). Information on sub-seismic-scale faults and fractures was derived for selected wells where image logs had been acquired (e.g. Figure 2.32b). A study was then performed on understanding the structural history of the site and the relationships between ancient and present-day stress fields and the observed faults and fractures. This work

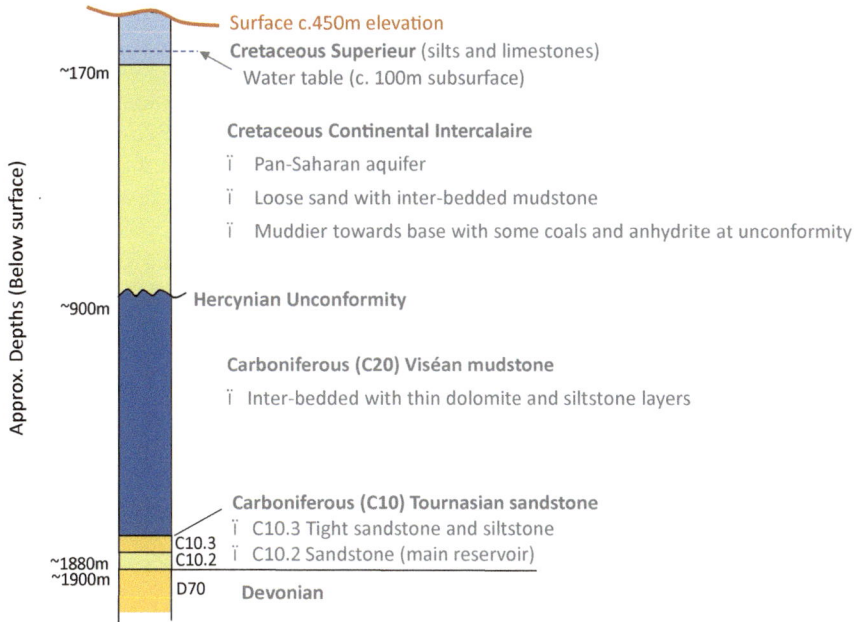

Fig. 2.31 Stratigraphic summary of the geology of the In Salah CO$_2$ storage site at Krechba, Algeria, where the CO$_2$ was injected into the 1.9 km deep Carboniferous sandstone unit (C10; Tournasian) beneath a thick mudstone sequence (C20 Viséan)

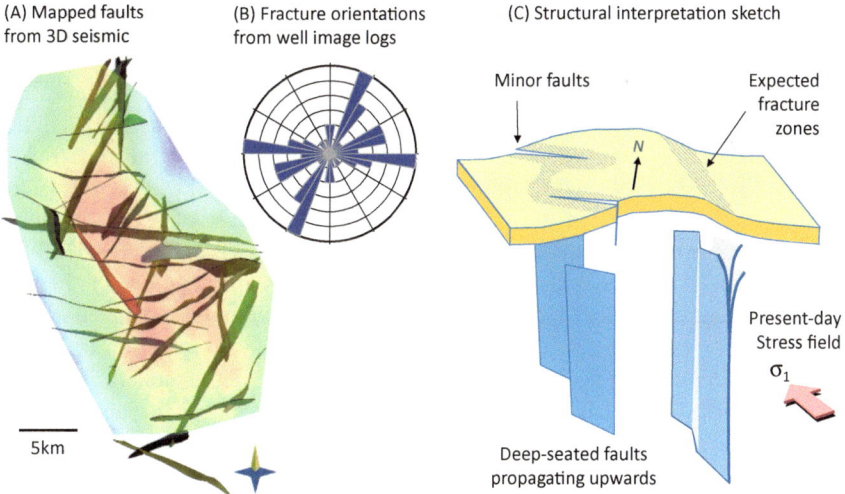

Fig. 2.32 Illustration of structural modelling activity for the In Salah case study: **a** birds-eye view of top reservoir surface with projections of fault surfaces from 3D seismic interpretation; **b** rose diagram of fracture plane orientations from image-log analysis; (C) Main elements of the structural interpretation and concepts

is described in Ringrose et al. (2009), Iding and Ringrose (2010), and Bond et al. (2013). The main steps in this work were:

1. To identify the various sets of faults and fractures and their relationships to the main tectonic events affecting this basin (summarized in Fig. 2.32c). The dominant groups of faults and fractures are the two sets aligned east-west and NE-SW. These were interpreted as having been mainly generated by strike-slip tectonic processes during the Hercynian orogeny (see Coward and Ries 2003). The gentle anticline forming the structural closure for the gas field and CO$_2$ storage unit was created by compressional tectonics (basin inversion) and influenced by deep-seated strike-slip faults.

2. To identify those faults and fractures which are critically aligned to the present-day stress field and therefore have potential to be hydraulically conductive or could be stimulated by elevated injection pressures. Since the maximum horizontal compressive stress is currently aligned at 135°N (Fig. 2.32c), critically-aligned fractures are expected to be vertical and either parallel to σ_1 (tensile mode) or $\pm30°$ from σ_1 (shear fractures), assuming a strike-slip regime.

Key features of this work are strain mapping using structural restoration and forward geomechanical modelling to understand fault and fracture failure modes. Two forms of strain were modelled: fault-related fractures and fold-related fractures (Bond et al. 2013). One important outcome of this study was that early CO$_2$ breakthrough observed in a monitoring well (KB-5) was due to fault-related and stress-aligned fractures.

The next major question addressed was to understand how these critically-aligned fractures control the observed dynamic flow behaviour (CO$_2$ breakthrough to well KB-5). Simple history-matching procedures can be used to adjust the flow simulation models to match observed pressures and flow rates. It is preferable, however, for these adjusted permeabilities to be based on independent observations. By using the detailed analyses of fracture datasets and by building high resolution fracture network models (Fig. 2.33), Iding and Ringrose (2010) made estimates of the likely fracture permeability at the scale of the full-field reservoir grid cells. Effective (fracture-network) permeabilities were found to be around 300 md (ranging between 140 md and 1000 md)—2 orders of magnitude higher than the matrix permeability (1–10 md). In order to achieve a history match to the dynamic observations about injection well KB-502, Iding and Ringrose (2010) applied factor of 1000 to represent the effect or the fracture-enhanced permeability corridor (Fig. 2.34). Similar conclusions were reached in reservoir modelling studies published by Bissell et al. (2011) and Shi et al. (2012), who also found that the fracture zone permeability must vary with time—that is, that the fracture zones are stress sensitive. These observations led to modifications of injection schedule at the KB-502 well, including temporary shut-ins and lowering of the injection pressure (Ringrose et al. 2013).

The third question addressed in this case study concerns understanding of the response of the rock system to CO$_2$ injection (Table 2.4). Bond et al. (2013) applied the knowledge gained from the near-well fracture-flow studies to generate a flow model for the whole Carboniferous Tournasian sandstone unit (ref. Figure 2.31).

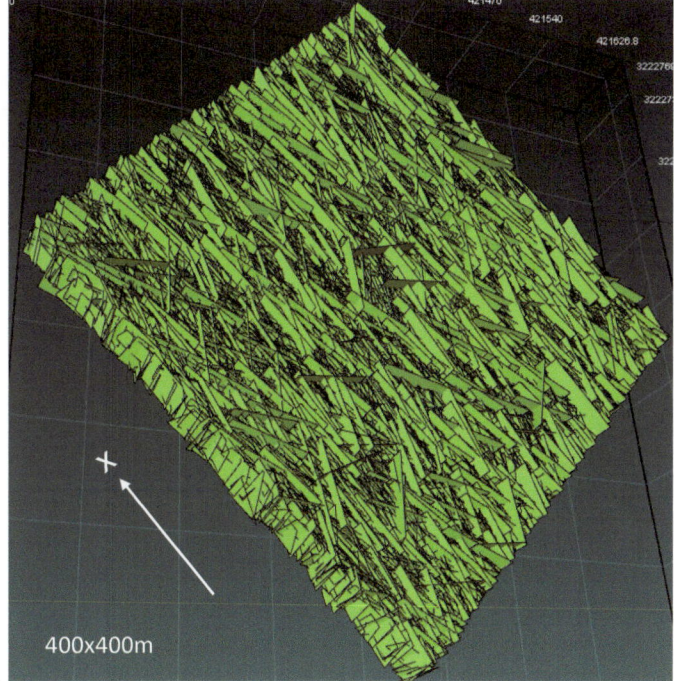

Fig. 2.33 Discrete Fracture Network (DFN) model of one largescale reservoir grid cell, constructed to estimate likely ranges of fracture permeabilities using geological data calibrated to well-test data (from Iding and Ringrose 2010; © Elsevier, reproduced with permission)

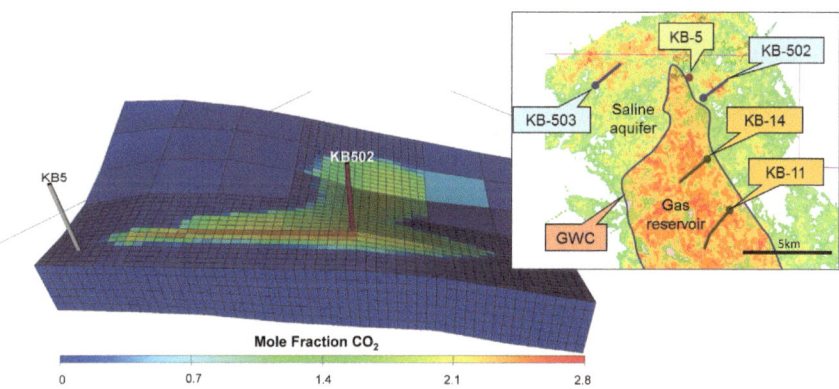

Fig. 2.34 Example reservoir simulation showing CO$_2$ distribution around injection well KB-502 with preferential flow along the fracture-enhanced permeability corridor leading to early breakthrough of CO$_2$ observed at well KB-5. Simulation shows CO$_2$ distribution at circa two years after injection start and immediately prior to breakthrough at KB-5. Inset shows well locations and scale

They estimated an anisotropic permeability tensor with a horizontal permeability ratio of $^2/_3$ governed by a stress-sensitive model (called fracture dilation tendency). Furthermore, the observed uplift pattern detected via analysis of Interferometric Satellite Airborne Radar (InSAR) data at this site (Mathieson et al. 2010; Vasco et al. 2010) shows a good match to the predicted anisotropic permeability tensor (Fig. 2.35). This confirms the hypothesis that the uplift pattern corresponds to the subsurface pressure field controlled by the permeability of the injection unit, as originally proposed by Vasco et al. (2008). Note that a very close match is observed for injection well KB-501 (indicating a pressure field mainly controlled by fractures) while a more approximate match is observed for KB-502 and KB-503 (due to the effects of specific faults influencing the actual pressure field). The geomechanical response and time-dependent flow properties of these fracture zones and faults have been extensively studied (e.g. Shi et al. 2012; Rinaldi and Rutqvist 2013; White et al. 2014).

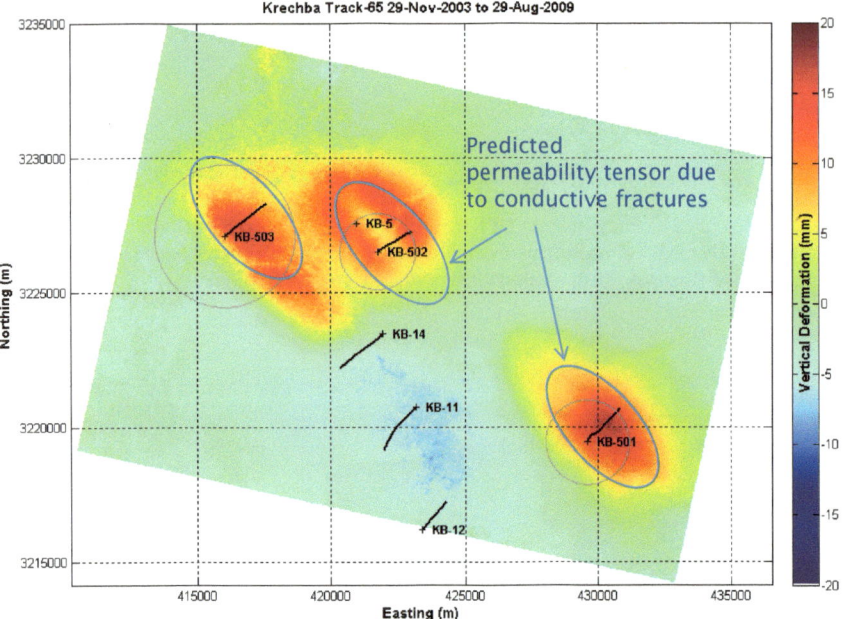

Fig. 2.35 Surface vertical deformation at Krechba (in August 2009) compared with permeability tensors estimated from fracture modelling by Bond et al. (2013). Satellite image based on processing of InSAR data by MDA/Pinnacle Technologies (redrawn from Mathieson et al. 2010); Horizontal CO₂ injection wells are KB-501, KB-502 and KB-503 and horizontal gas production wells are KB-11, KB-12 and KB-14. Grey circles indicate relative CO₂ mass injected per well, scaled to 3Mt injected in all three wells by this time

2.5.4 Modelling CO₂ Storage at Sleipner

The long-running Sleipner CCS project, which started injecting CO_2 into a deep offshore saline aquifer formation in 1996, represents the most important case study for understanding CO_2 storage flow dynamics. Sleipner is located offshore Norway and utilizes the world's first offshore platform-based CO_2 amine capture facility (Torp and Gale 2004; Hansen et al. 2005). The CO_2 is stored in the Utsira formation at a depth of 800–1000 m below the sea surface (water depth is 82 m). Here the Utsira Fm. comprises a 200–300 m thick sandstone unit, with several thin shales. The formation water is saline (salinity close to seawater), and the sandstone has high porosity (38%) and permeability (1–8 Darcy). The Utsira Fm. is overlain by a thick shale caprock, with some silty intervals in the upper part. One shallow long-reach highly deviated well is used to take the CO_2 far enough away (2.4 km) from the producing wells (to avoid any external corrosion of casing) and from the production platform (Baklid et al. 1996).

The monitoring plan for this injection site is dominated by the use of time-lapse seismic surveys (Chadwick et al. 2010; Eiken et al. 2011; Furre et al. 2017), which as well as providing assurance of safe and effective storage to the stakeholders (government bodies, site operator, and the public) also gave excellent imaging of the plume allowing flow models to be calibrated and the flow processes to be better understood (Fig. 2.36).

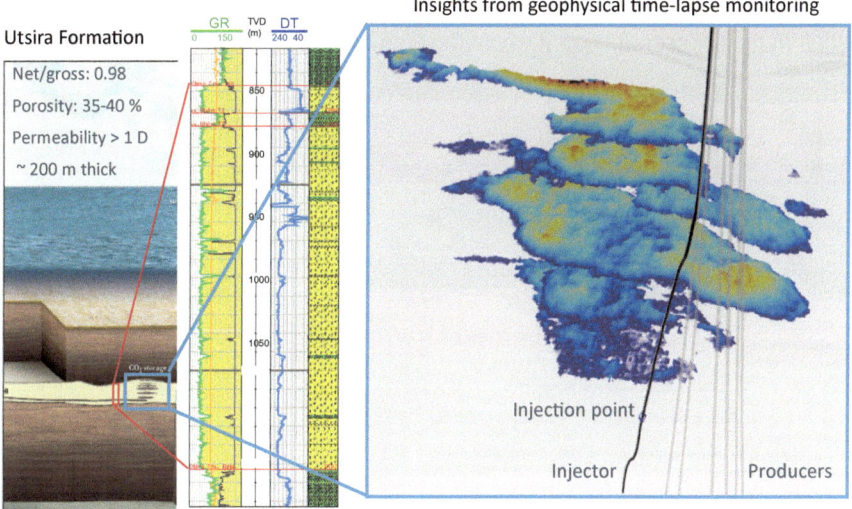

Fig. 2.36 Site summary of the Utsira sandstone formation at the Sleipner injection site, with example well logs (GR = gamma ray; DT = sonic) and a 3D reconstruction of time-lapse seismic imaging data revealing the distribution of the multi-layer CO_2 plume in 2010 (using an interpolation method developed by Kiær et al. 2016). (Images courtesy of Equinor)

The seismic imaging datasets allow us to estimate the lateral extent of the CO_2 plume, given several important constraints. The large contrast in acoustic properties between the in situ brine and the injected CO_2 gives a strong time-lapse seismic response (ref. Fig. 2.24). However, there are several limits to the detection of CO_2 from seismic-amplitude difference data. Under optimal conditions, such as for the uppermost layer 9, CO_2 layer thicknesses down to the meter scale can be observed (Furre et al. 2015), but in deeper layers signal degradation occurs due to inelastic attenuation and transmission loss caused by the overlying CO_2 layers (Boait et al. 2012). Despite these limitations, good imaging and mapping of the multi-layer CO_2 plume can be achieved at Sleipner.

There have been numerous studies on modelling various aspects of the CO_2 plume at Sleipner (e.g. Zweigel et al. 2004; Bickle et al. 2007; Chadwick and Noy 2010; Singh et al. 2010; Cavanagh 2013; Williams and Chadwick 2017). Our aim here is not to compare the different approaches in detail, but rather to illustrate the knowledge gained about flow dynamics from these studies. Figure 2.37 summarizes the geological and seismic information which is available to constrain the dynamic models. An important point here is that only the top and bottom of the Utsira sandstone and the top of the over lying sand wedge were visible for seismic data prior to injection. Numerous thin shales within the sandstone unit were evident from well data (gamma log; Fig. 2.36), but their effect of the vertical migration of CO_2 within the sandstone was unknown before injection started. As repeat seismic surveys were acquired and interpreted, the multi-layer character of the plume became apparent (Fig. 2.37), stimulating repeated efforts to better match the observations with models.

Many of the efforts have focused on successful matching of the areal plume distribution in the uppermost layer 9 where the high quality of seismic imaging provides

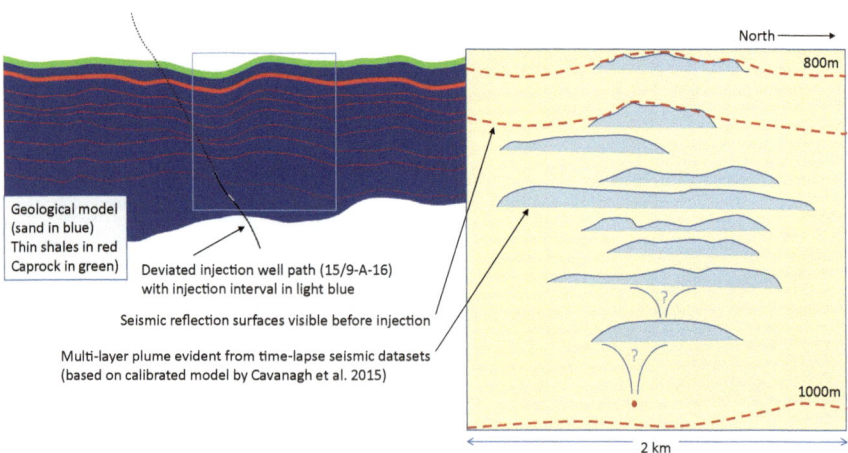

Fig. 2.37 Summary of information available for modelling the multi-layer CO_2 plume at Sleipner. The nine CO_2-filled layers are only evident from the time-lapse seismic data and have uncertain geometry and connection points between layers

the best basis for forecasting. Singh et al. (2010) also published a reference dataset with available estimates (and ranges) rock and fluid properties, as a frame for modelling. More recently, an updated and extended reference model has been released (Andersen et al. 2018). Figure 2.38 shows some selected examples of dynamic simulations of the layer 9 plume. Note how earlier models (e.g. Chadwick and Noy 2010) predicted a more circular plume geometry, but that later studies were able to capture the elongated shape, partly by better inclusion of the effects of gravity forces in the simulation (Cavanagh 2013) but also by modifying rock properties to match the effects of geological architecture (Williams and Chadwick 2017). The CO$_2$ distribution in layer 9 is observed to closely match the topography of the top structure (evidence of the dominance of fluid buoyancy forces), but also to align with north-south trending channel features evident from the seismic data.

Modelling the full multi-layer plume is more challenging due to uncertainties in both the geology (especially location of flow pathways through the thin shales) and the fluid properties (especially because the in situ temperature is poorly constrained). Figure 2.39 shows an example 3D simulation of the plume at Sleipner, confirming that FD flow simulation models can reproduce the stack of multiple relatively-thin CO$_2$ filled layers observed on seismic. An alternative modelling approach which works well for the Sleipner case is Invasion Percolation (IP). The approach neglects viscous forces and assumes that the buoyant non-wetting phase fills available traps,

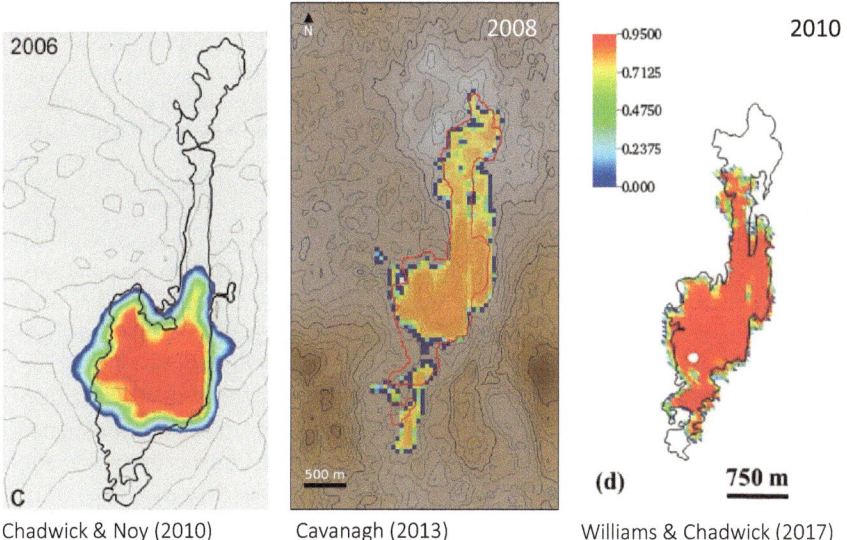

Fig. 2.38 Examples of dynamic simulations aimed at matching observed plume (black or red outlines) in the uppermost layer nine where seismic detection is most confident. Each of these cases use multi-phase FD simulators with different assumptions (see references for details). The date at the top of each image is the seismic survey date for the model history-match. Images reproduced with permission, © Geological Society of London (left) and © Elsevier (centre and right)

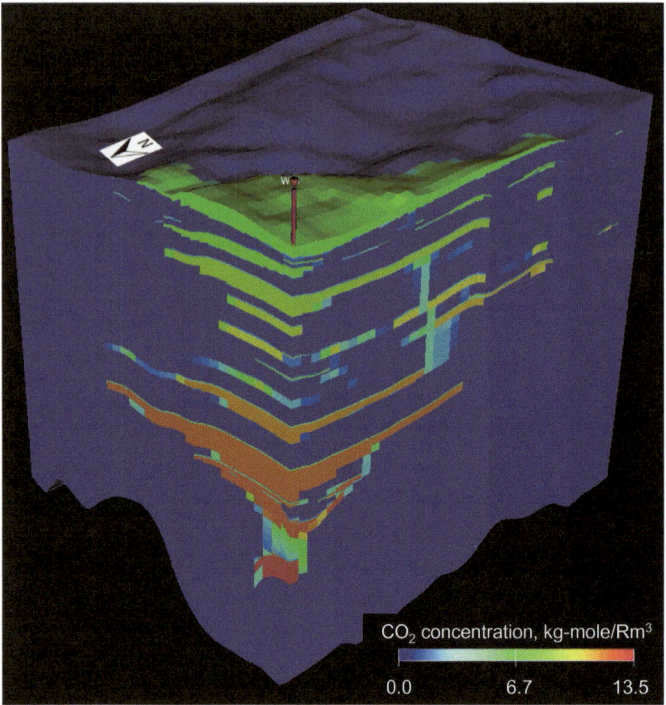

Fig. 2.39 Example multi-phase flow simulation model of the multilayer CO$_2$ plume at Sleipner showing a forecast to 30 years of injecting the CO$_2$-rich stream. This view of the 3D simulation grid intersects the centre of the plume at the injection point showing the lateral extent of plume in each layer. The CO$_2$ concentration is given in kg-mole/Rm^3. Z scale is exaggerated by a 7. From Nazarian et al. 2013; © Elsevier, reproduced with permission

defined by the capillary threshold pressure field. Figure 2.40 shows an example from Cavanagh and Haszeldine (2014), where a good overall match to observations at year 2002 was achieved. However, the authors found that the IP simulation match required very low threshold pressures for the shale barriers. Recently, Furre et al (2019) have shown the importance of several vertical connection pathways in controlling flow between the CO$_2$-filled layers. These new insights will no doubt lead to further improvements in model matching at Sleipner. Nevertheless, this experience of developing dynamic fluid simulation models calibrated to seismic observations has led to many improvements in the understanding of the flow dynamics of CO$_2$ storage, generally showing the dominant role of gravity-dominated flow processes. This case study thus provides an important foundation for modelling storage at future sites—flow models calibrated against the Sleipner case should provide a sound basis for forecasting elsewhere.

Table 2.5 gives some summary CO$_2$ storage metrics for the Sleipner case, as of 2010, including the estimate of the storage efficiency factor, ε, discussed above. Differentiating between the free phase CO$_2$ and the portion dissolved in the brine

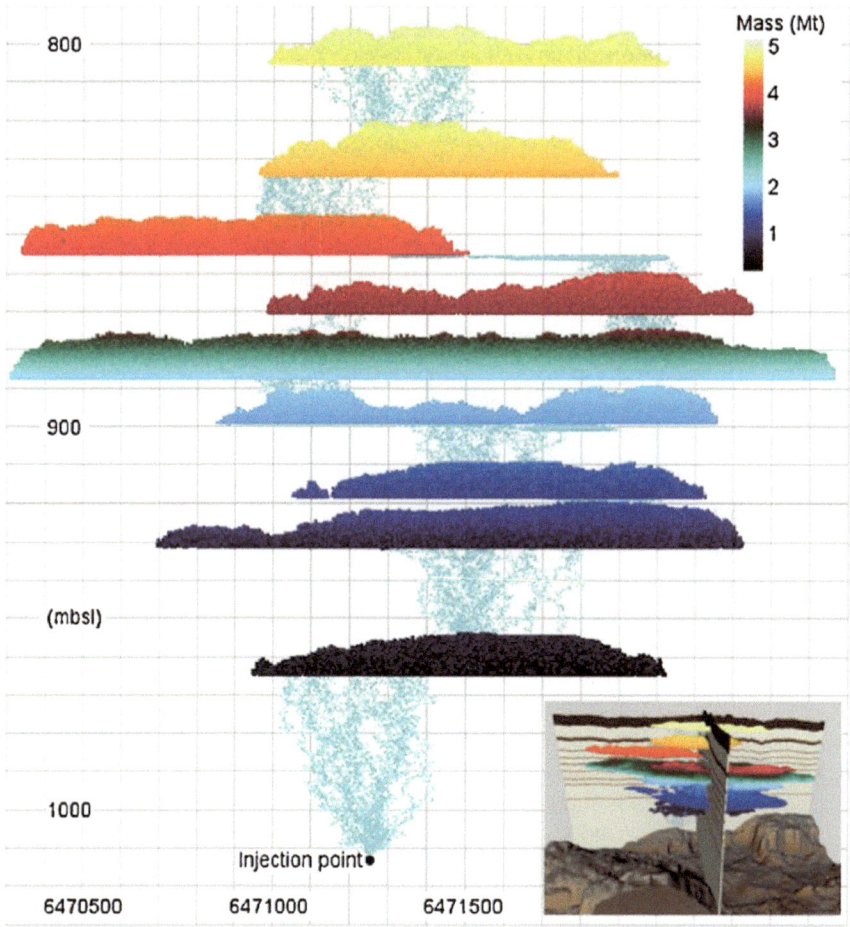

Fig. 2.40 Example Invasion Percolation (IP) simulation for the Sleipner storage site (from Cavanagh and Haszeldine 2014). Result is for 5 Mt of CO$_2$ injected by July 2002. Light blue lines are CO$_2$ migration pathways. Colours indicate filling sequence and mass

Table 2.5 CO$_2$ storage metrics for the Sleipner case as of 2010

CO$_2$ storage metric (as of 2010 seismic survey)	Mass (Mt)	Fraction of pore space occupied (ε)
Total injected	12.18	0.048
Free phase	11 ± 0.5	0.044
Dissolved phase	1.2 ± 0.5	0.004

phase is more challenging. Most of the free-phase CO_2 is expected to be mobile but retained beneath gentle structural closures beneath thin shales. A smaller fraction may be residually trapped due to migration, but there is no direct observation of this fraction at Sleipner. Using time-lapse gravity data (giving density change as a function of time) combined with the time-lapse seismic data (giving estimates of geometry and volumes), Alnes et al. (2011) showed that the rate of CO_2 dissolution into the brine at Sleipner must be less than 1.8% per year. However, the dense brine cannot be directly detected from the gravity data, so it could be much lower than this upper bound. Laboratory and modelling studies can be used to argue that some dissolution must, however, be occurring. A recent model forecast for layer 9 by Cavanagh et al. (2015) suggests 10% dissolution after 20 years injection at Sleipner. This (albeit very uncertain) estimate gives a likely picture for the fractionation between free-phase and dissolved-phase CO_2 at Sleipner (Table 2.5).

2.6 Calculating Injectivity

2.6.1 Concepts and Methods

Once we have established a reasonable expectation of the required storage capacity for a project, we need to start thinking about designing and managing the injection wells and transport infrastructure. Figure 2.41 gives a simple overview of the main engineering aspects of the transport and storage system especially related to pressure management. We need to think about:

- CO_2 supply—rates, pressures and temperatures
- Reservoir depth, water depth
- Well design
- Site performance (plume behaviour)
- Reservoir properties
- Overburden and seal characteristics
- Effects of regional aquifers.

For the well, there are two main pressure points to consider—the wellhead pressure, P_{wh}, and the bottom-hole pressure, P_{bh}. Then there are two main pressure gradients to think about: in the wellbore and from the well into the formation (Fig. 2.41).

Under static conditions, the bottom-hole well pressure, P_{bh}, can be simply estimated from the wellhead pressure, P_{wh}, by:

$$P_{bh} = P_{wh} + \rho_{CO_2} g \, \Delta h \qquad (2.29)$$

However, under dynamic conditions, with varying flow rates, more advanced methods are needed with a good appreciation of the CO_2 phase behaviour. We will focus here on understanding the injectivity of a CO_2 injection well, as this is one of

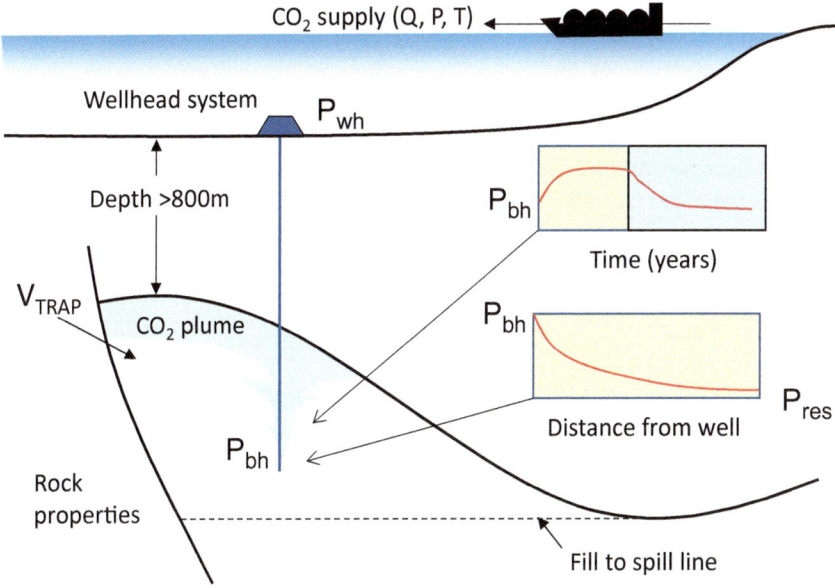

Fig. 2.41 Summary of issues for injection pressure management—based on an offshore setting

the key questions for site assessment (alongside the storage capacity and containment themes). The expected injectivity for a CO_2 injection well is related to three main factors:

(a) the well design itself,
(b) the well placement strategy, including well angle and completion length, and
(c) the reservoir formation properties, especially the permeability.

In its simplest form, the Injectivity index, II, for a water well is given by:

$$II = \frac{q}{(p_{fbhp} - p_{res})} \qquad (2.30)$$

where q is the flow rate, p_{fbhp} is the flowing bottom-hole pressure and p_{res} is the far-field reservoir formation pressure. This assumes the injection fluid is incompressible. For a gas injection well, II_{gas} may be estimated using the p^2 technique, valid for low pressure gas injection (Lee and Wattenbarger 1996):

$$II_g = \frac{q}{(p_{fbhp}^2 - p_{res}^2)} \qquad (2.31)$$

For a reservoir injection unit with known reservoir properties the flow rate, q, can also be estimated using a radial Darcy flow equation (assuming a vertical well geometry):

$$q = \frac{2\pi k_{res} h_i (p_{res} - p_{fbhp})}{\mu \ln(r_e/r_w)} \tag{2.32}$$

where k_{res} is the permeability of the rock formation, h_i is the height of the injection well interval (the completion interval), μ is the fluid viscosity, r_e is the effective radius of the reservoir unit and r_w is the radius of the well itself. Golan and Whitson (1991) showed that for high flow rates this equation may be adapted and applied for gas production wells, giving:

$$q_g = \frac{1.406\, k_{res} h_i \left(p/\mu_g Z\right)(p_{res} - p_{wi})}{T[\ln(r_e/r_w) - 0.75 + s + Dq_g]} \tag{2.33}$$

where $p/\mu_g Z$ is the pressure depth function (assumed to be linear), T is temperature, s is the skin factor and Dq_g is a rate dependent skin factor. Rearranging Eq. 2.33 and neglecting the skin factors, we can obtain a function for estimating the injectivity of a CO$_2$ injection well assuming high flow rates:

$$II_{CO_2} = \frac{q_g}{(p_{wi} - p_{res})} = \frac{1.406\, k_{res}\, h_i \left(p/\mu_g Z\right)}{T[\ln(r_e/r_w) - 0.75]} \tag{2.34}$$

For lower flow rates Eq. 2.31 is more appropriate, and a more general method using pseudo pressure functions (Al-Hussainy et al. 1966) may also be required for more complex cases. In practice, the calculation of q_{CO2} or II_{CO2} is complicated by many factors, including the variations in CO$_2$ phase density within the wellbore, multiphase flow effects, near-wellbore damage and heterogeneity within the rock formation. As a result, CO$_2$ storage screening studies tend to use rather simple, and at times over-optimistic, models for injectivity.

2.6.2 Example Injectivity Challenges

To illustrate issues with injectivity, we can use data from the first years of CO$_2$ injection at the Sleipner project (Fig. 2.42). Despite a very successful overall project history, the initial stages of the project were marked by some technical challenges with regard to the stability of the injection (Hansen et al. 2005; Ringrose et al. 2017). After starting injection in September 1996, the planned injection (using a 100 m perforation interval in a horizontal well at a depth of 1014 m) did not initially achieve the intended injection rates due to sand influx into the well. Installation of sand screens in December 1996 partially improved the injection rates; however, significant improvements in injectivity were only achieved following re-perforation of the injection interval in August 1997 and by installation of a gravel pack and sand screens over a 38 m interval of the horizontal section. Following this well work-over operation, stable injection rates of 1.4–1.6 MSm^3/day (or 2600–3000 tonnes/day) were achieved (Fig. 2.42) allowing handling of the intended injection

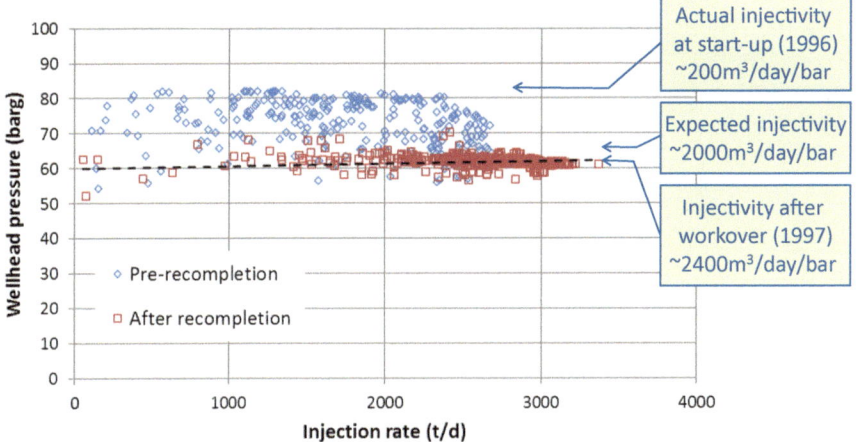

Fig. 2.42 Sleipner injection data (1996–1999)

rates. Figure 2.42 also illustrates how higher injection wellhead pressures (c. 80 bar) were required prior to the well work-over operation, reducing to 62–65 bar for the subsequent period. In terms of injectivity, the pre-injection injectivity was expected to be around ~2000 m³/day/bar (using Eq. 2.34), while the actual injectivity levels dropped down to around 200 m³/day/bar initially and then subsequently returned to around 2000–2400 m³/day/bar after the well re-perforation operation. It should also be noted that the Sleipner well injects CO_2 volumes well below the well injectivity limits, and that well injectivity limitations were only experienced in the first 6 months.

This example of early-stage problems with injectivity of the well at Sleipner is part of the general class of problems involving near wellbore resistance to injection, generally known as the 'skin' effect. That is, some form of damage to the natural reservoir formation that causes reduced permeability. Typical causes of near-wellbore damage include: invasion of the formation with mud filtrates, local collapse of weakly consolidated sandstone and migration of fine particles into the pore space. When drilling the well, efforts are made to minimize these effects, but that is not always possible. Well completion technology (setting of perforations shots, installing liners, etc.) is used to ensure the possible effects of near-wellbore damage are optimally handled and mitigated.

These effects are simply illustrated in Fig. 2.43. Here we modify Eq. 2.30 to include a resistance term ΔP_{skin} to represent additional resistance due to wellbore damage:

$$II = q/(P_{wf} - P_{res} - \Delta P_{skin}) \tag{2.35}$$

The dotted line on Fig. 2.43 represents the additional pressure drop due to wellbore damage, while the solid line shown what would be expected if there were no damage effects. The horizontal axis is between the well bore radius, r_w, and the effective radius

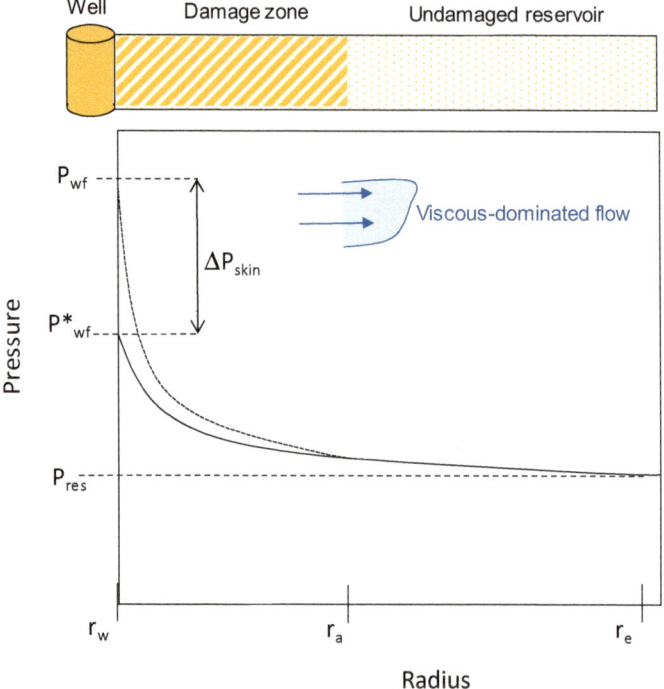

Fig. 2.43 Illustration of effect of near-well damages zone on pressure gradients around an injection well

of the reservoir, r_e, with the radius of the damage zone, r_d, in between. Working out what is really happening in the near-wellbore region is actually very difficult in practice, but the concepts involved are important. What is generally observed is that the bottom-well flowing pressure P_{wh}, is higher than the expected well pressure P^*_{wh} (as experienced at Sleipner in the early stages). It should be noted that it is also possible to have 'negative skin' where induced fractures in the near-well region cause less resistance to flow than expected.

After the near-well flow resistance issues at Sleipner were resolved and mitigated (using a gravel pack and sand screens), injection continued with very little formation resistance due to the very high permeability of >1 Darcy. In fact, the pressure difference between the bottom-hole pressure (not directly measured) and the formation pressure is thought to be less than 2 bars (Eiken et al. 2011) and the well injects at rates well below the injectivity limit. In such a case the pressure profile might look more like Fig. 2.44. These near-well effects will also influence the CO$_2$ distribution in the reservoir (illustrated by the insets on Figs. 2.43 and 2.44). High formation resistance, and consequently high-pressure gradients, will lead to a more viscous-dominated flow regime, while very small pressure gradients close to the well will mean that gravity forces will dominate even close to the well.

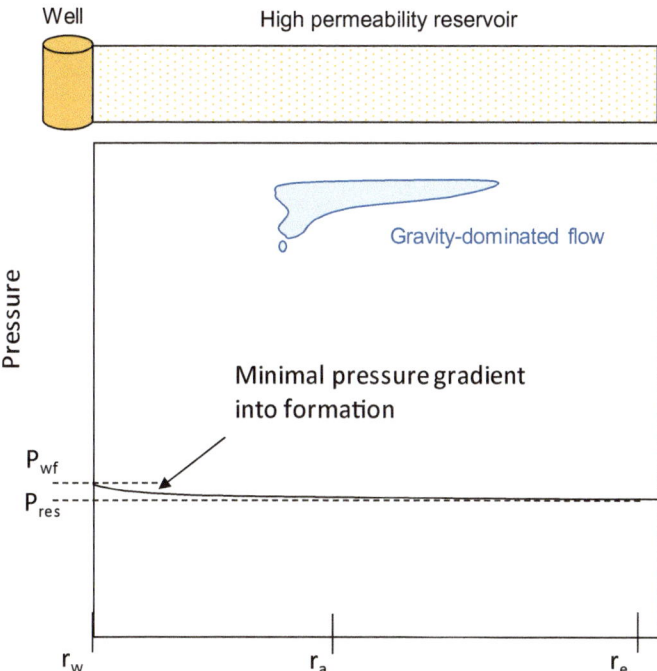

Fig. 2.44 Illustration of effect of pressure gradients around an injection well for high-permeability reservoir with no significant well-bore damage effects

Another important phenomenon which can lead to significant loss of injectivity in a CO_2 injector is the effect of salt precipitation. Salt precipitation can be triggered by the evaporation of brine at the interface with the injected CO_2 stream, and this is especially the case if the CO_2 stream has low aqueous content (i.e. very dry CO_2). The salt crystals can partially block the pore space and lead to a decline in injectivity (Miri et al. 2015). The effect is expected to occur close to the well (Fig. 2.27) and was experienced in the early stages of injection at the Snøhvit site (Fig. 2.45) (Grude et al. 2014; Pawar et al. 2015) and was also observed at the Ketzin CO_2 injection pilot project (Baumann et al. 2014). The solution chosen to mitigate or avoid this problem at Snøhvit was to inject episodic slugs a Methyl–ethylene–glycol (MEG) solution to the injection stream (Hansen et al. 2013). As a result, most CO_2 injection projects need to evaluate whether there will be a need to include salt precipitation inhibitor solution in the design—a conclusion that depends on the chemistry of the brine and CO_2 stream.

The longer-term rising pressure trend at Snøhvit was caused by the presence of geological barriers, which impeded pressure dissipation into the reservoir unit as a whole (interval 'b' on Fig. 2.45). This impedance to flow in the far-field reservoir (the main barrier was inferred to be about 3000 m from the well; Hansen et al. 2013), led to the decision to modify the injection in order to access a better reservoir unit

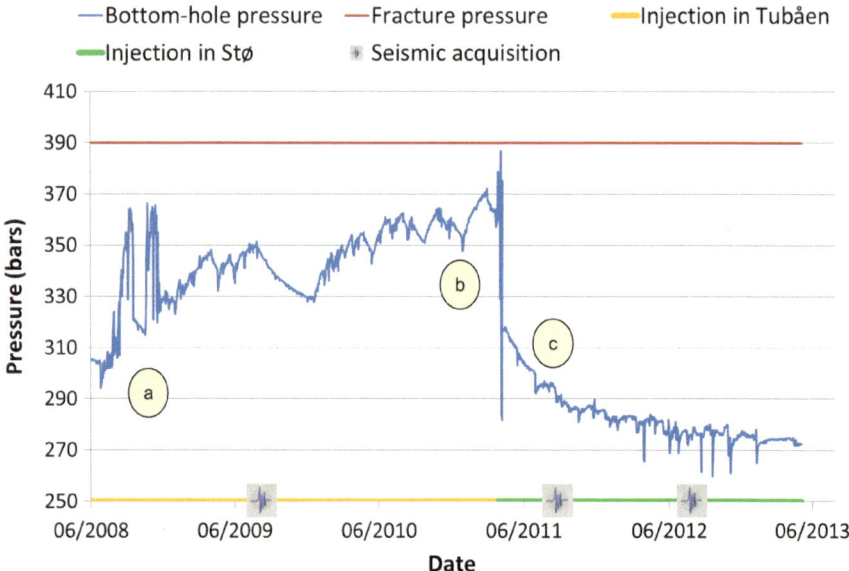

Fig. 2.45 Pressure history at the Snøhvit CO$_2$ storage site (2008–2013) with time-lapse seismic acquisition surveys. Three main features of the injection pressure history are: **a** early rise in pressure due to near-wellbore effects related to salt drop-out, **b** a gradual rising trend in pressure due to geological flow barriers in the Tubåen Fm., and **c** pressure decline to a new stable level following well intervention and diversion of the injection into the overlying Stø Fm

higher up using the same injection well (Hansen et al. 2013; Pawar et al 2015). The Snøhvit (Tubåen reservoir) injection history thus provides a useful illustration of how resistance to injection in both the near-well and far-well regions was handled and mitigated.

Each injection well will have unique features related to the formation properties, the fluid system and the reservoir heterogeneities. However, by comparing example well performance metrics (e.g. Figure 2.42) and drawing on the lessons learned from these early projects we can at least understand the likely range in behaviour. Well performance is ultimately an engineering design problem, and as the technology of CO$_2$ storage progresses we can expect to see further optimization of well designs and the development of improved forecasts of injection rates and associated uncertainties for future projects.

2.7 Geomechanics of CO_2 Storage

2.7.1 Understanding Stress and Strain

The topic of the geomechanical response to CO_2 injection has drawn a lot of interest—with concerns often being raised about the likelihood of rock failure and possibility for induced seismicity. To develop an informed approach to these issues, it is helpful to bear in mind some over-arching principles about rock stress and fluid pressures. Here we will focus only on the situation in sedimentary basins where we expect most CO_2 storage projects to be located (rather than in continental basements). Figure 2.46 illustrates the underlying principles:

- Rock stress is mainly controlled by the overburden weight, σ_v (which can be estimated from rock density), which in extensional basins is usually equal to the maximum stress vector, σ_1. The other stress components, σ_2 and σ_3, are determined by the far-field tectonic stresses and the rock strength. However, in strike-slip tectonic regimes the intermediate stress σ_2 vector is vertical and in thrust tectonic regimes it is the minimum stress, σ_3, that is vertical.

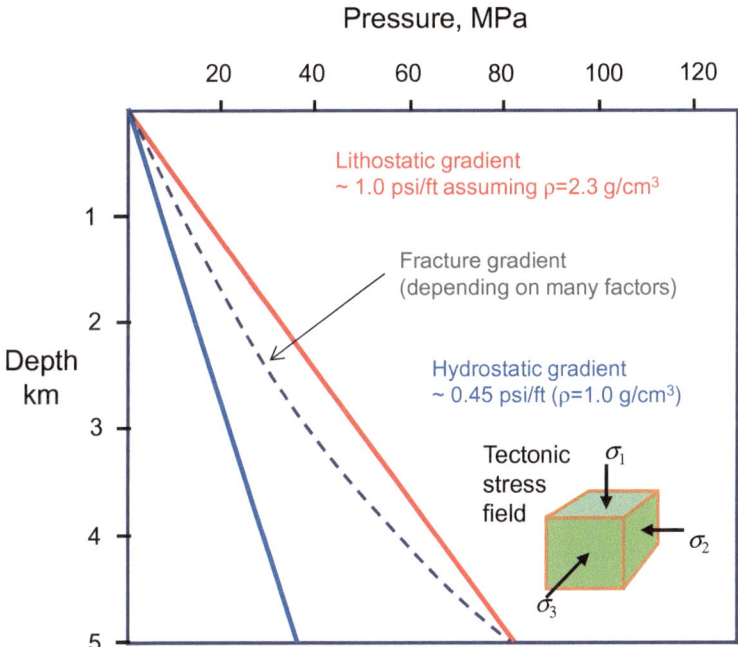

Fig. 2.46 Simple sketch of rock stress and fluid pressure in a sedimentary basin (gradients calculated using simple uniform properties)

- In the shallower parts of sedimentary basins, the fluid pressure is usually in hydro-static equilibrium, meaning the pressure is equal to the weight of water beneath a reference point close to the earth's surface (i.e. sea level or the water-table level). However, at some depth a situation of overpressure can occur—meaning pressures can become significantly higher than hydrostatic.
- The point at which rocks fracture, often called the fracture gradient, is related to the minimum stress, σ_3, but also to the depth in the basin. As rocks get hotter with depth there comes a point where the fracture pressure gets very close to the maximum stress value (because rocks become less rigid and more plastic at these depths).

A key objective in a CO$_2$ storage project is therefore to ensure that the injection pressure does not exceed the fracture pressure. This is an apparently simple objec-tive—although it can be quite complicated to determine exactly what the limiting pressure should be. Most CO$_2$ injection projects are likely to target the depth interval of 1–4 km. Projects need to be deeper than 800 m to ensure CO$_2$ is in the dense phase, and at depths deeper than about 4 km the rock properties are often too low for high rate injection to be achievable.

We also need to understand (or remind ourselves of) some essential rock mechan-ical concepts:

1. **Stress and strain:** The imposed stress field, explained briefly above, results in some degree of rock deformation, called strain (Fig. 2.47). Stress is related to strain via Young's Modulus, E, such that:

$$\underline{\sigma} = E\underline{\varepsilon} \tag{2.36}$$

   where $\underline{\sigma}$ and $\underline{\varepsilon}$ are the mean stress and strain.
   Strain can be resolved into three orthogonal components ($\varepsilon_1, \varepsilon_2, \varepsilon_3$) corresponding to the triaxial stress field, or can be described by a full tensor. Strain can also be expressed as a volumetric strain, ε_V.

2. **Effective stress**: This is the net stress operating on the granular rock framework and is defined by:

$$\sigma_{eff} = \sigma - p \tag{2.37}$$

Fig. 2.47 Simple sketch of triaxial rock stress field, rock strain and pore pressure

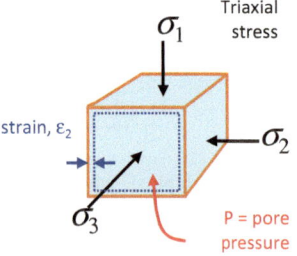

where σ is the total stress and p is the pore pressure. In practice, we often need to resolve the stress tensor into its components, e.g. horizontal and vertical or minimum and maximum.

3. **Rock compressibility** is a measure of volume change as a function of pressure (or mean stress). The bulk rock compressibility, c_r, can be defined as follows (for isothermal conditions):

$$c_r = -\frac{1}{V_p}\left(\frac{dV_p}{dp}\right)_T \tag{2.38}$$

Typical values for rock compressibility are in the range of 10^{-11} to 10^{-9} Pa^{-1} (that is to say, rock is not very compressible).

4. **CO$_2$ compressibility** which depends on the in situ pressure and temperature conditions, can be usefully expressed as a function of fluid density (Vilarrasa et al. 2010):

$$c_f = \frac{1}{\rho_f}\left(\frac{d\rho_f}{dp_f}\right) \tag{2.39}$$

- CO$_2$ compressibility values (at subsurface conditions) lie in the range of 10^{-9} to 10^{-8} Pa^{-1} (that is to say, ~2 orders of magnitude higher than the rock compressibility).

For the CO$_2$ storage problem, it is useful to link the compressibility terms with the effective stress equation for a porous medium. Nordbotten and Celia (2012) define the compressibility for a porous medium, C_ϕ, as follows:

$$c_\phi = -\frac{d\phi}{d\sigma_{eff}} = \frac{d\phi}{dp} \tag{2.40}$$

This assumes that porosity changes as a function of effective stress and pressure are equivalent.

By combining this relationship with a mass balance equation for a single-phase pore fluid with known density, Nordbotten and Celia (2012) go on to explain that:

$$\rho(c_\phi + \phi c_f)\frac{dp}{dt} = \rho c_\Sigma \frac{dp}{dt} \tag{2.41}$$

where $c_\Sigma = c_\phi + \phi c_f$.

This means that the total compressibility coefficient for the rock-fluid system (which is unknown) can be estimated from the fluid and rock compressibility terms (which can be measured or estimated).

We will refer to these basic concepts when discussing the key issues for CO$_2$ storage projects below. For further discussion and appreciation of these topics in rock mechanics, refer to fuller treatments such as Fjær et al. (2008) or Zoback (2007).

There are two main concerns often discussed in relation to rock mechanical aspects of CO_2 injection projects:

- Is there enough room for the CO_2 to be stored?
- Will the injection cause earthquakes?

In this short treatment we will try to set the framework for addressing these questions, with a view to avoiding some misunderstandings which have clouded the discussion.

A useful starting point for the question of the available room for CO_2 storage, is the analysis by Ehlig-Economides and Economides (2010) who published a paper on 'Sequestering carbon dioxide in a closed underground volume' where they concluded that:

> ... the volume of liquid or supercritical CO_2 to be disposed cannot exceed more than about 1% of pore space. [And that this] renders geologic sequestration of CO_2 a profoundly non-feasible option for the management of CO_2 emissions.

This study argued that the maximum storage efficiency value, ε, was around 1%—significantly lower than the values used by CO_2 storage capacity estimation studies (ref Sect. 2.4.4 above). There were many reactions to this paper, with one published response by Cavanagh et al. (2010) who argued that the analysis was flawed and based on an incorrect conceptual model and an overly simplistic mathematical analysis.

So, is it the case that only 1% of the storage pore volume is available for storage? In fact, most researchers agree that you cannot inject much fluid into a 'confined box' (the argument that Ehlig-Economides and Economides 2010 were making), but the issues are wider than that.

To understand this problem, we need to appreciate that there are three essential limiting factors:

1. The size of the box (the storage unit)
2. The properties of the box boundaries (faults and shale sealing units)
3. The ability of the box to absorb increased pressure (rock and fluid compressibility).

Zhou et al. (2008) published a study of the storage limits for a range of systems (Fig. 2.48a) and concluded that storage efficiency is around 0.5% for perfectly closed systems, but that a semi-closed system with a seal permeability of 10^{-17} m^2 (0.01 md) or greater behaves essentially like an open system with respect to pressure build up (due to brine leakage). Recall that for open systems, we expect storage efficiency values of around 4–6% (Sect. 2.4.4 above). From a geological perspective, although fault blocks in sedimentary basins can create confined pressure compartments, these are unlikely to be 'perfectly-sealed' since faults are complex zones of rock deformation and will have some permeability (generally very low but not zero). A similar argument holds for sealing layers like shales which have low but non-zero permeability. Furthermore, 3D fault architecture at the basin scale is likely to lead to some points of pressure communication (Fig. 2.48b) through zones with lower fault displacement or fault zones with sand-to-sand juxtapositions. Thus, the argument that

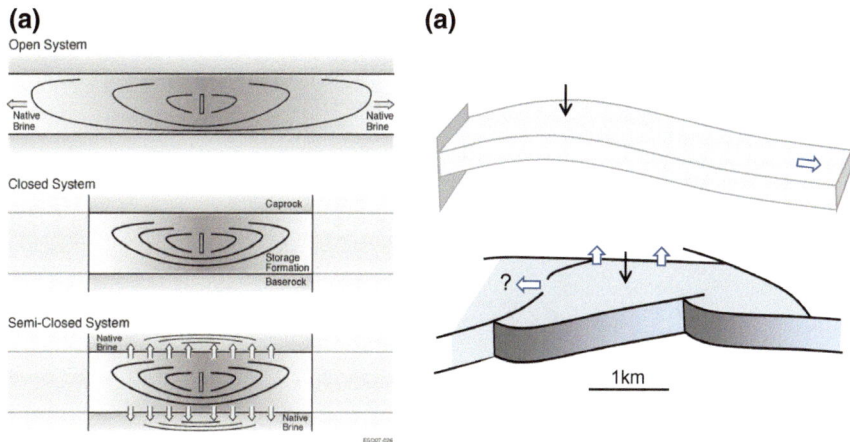

Fig. 2.48 Sketches of **a** open, closed, or semi-closed systems around an injection well interval (Image from Zhou et al. 2008). **b** Example 3D geometries of semi-open and semi-closed geologic storage systems (black arrows indicate injection points; white arrows indicate pressure dissipation routes). Lefthand image © Elsevier, reproduced with permission)

only 1% of the pore volume is available for storage is based on the very limiting assumption of a confined box. Nevertheless, the 'end member case' of injection into a sealed box can be a useful case to consider. In this limiting case, the available storage volume is a simple function of rock compressibility and the size of the aquifer unit / fault block (as summarized in Fig. 2.49). Within this framework, small fault compartments are unlikely to provide good storage targets, but as long as the storage unit is fairly large (>10 km), pressure dissipation is likely to be sufficient to allow storage to proceed without hitting the pressure limit.

2.7.2 CO₂ Storage and Induced Seismicity

Could CO₂ injection cause earthquakes? This is a much-debated issue—which again is clouded by some misunderstandings. In general, injecting water into, or close to, faults does reduce the effective stress and can lead to enhanced fault slip. However, injecting CO₂ is not the same as injecting water. Water is close to incompressible, while CO₂ is much more compressible. Furthermore, CO₂ is a non-wetting phase, tending to displace the water phase and enter the largest pores. These differences are borne out from observations from early CO₂ injection projects—there may be some induced seismicity, but the relationships are complex, and the level of induced seismicity is generally very low. Verdon et al. (2013) report findings for three injection sites, including the Weyburn project with a long injection history, which reveals complicated patterns of micro-seismicity. Most microseismic events were related to

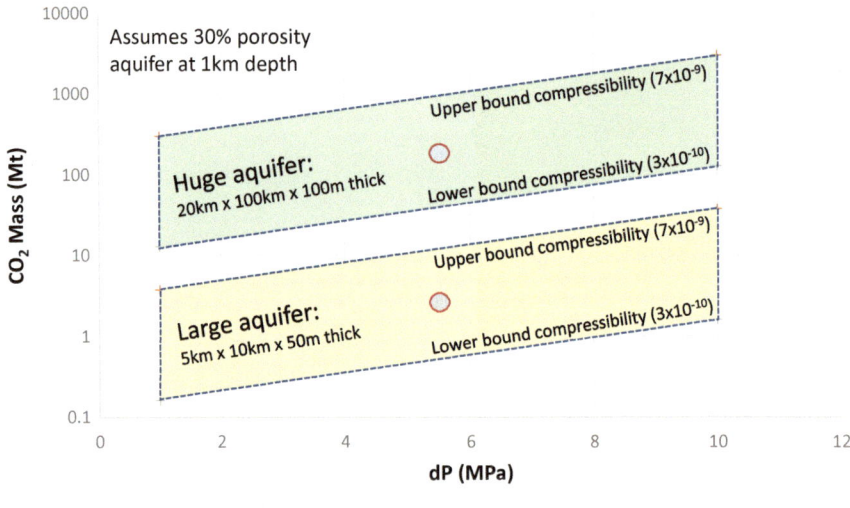

Fig. 2.49 Example storage volume estimates for selected closed-aquifer fault blocks based on compressibility estimates

production wells or occurred during shut-in of CO_2 injection wells and not during injection.

In a paper published in 2012, Zoback and Gorelick suggested that large-scale CO_2 storage could lead to large earthquakes, leading the authors to conclude that "large-scale CCS could be a risky, and likely unsuccessful, strategy for significantly reducing greenhouse gas emissions." However, their analysis was focused on the case of injection into brittle rocks commonly found in continental interiors. In responding to the claim made by Zoback and Gorelick (2012), Vilarrasa and Carrera (2015) argued that "Geologic carbon storage is unlikely to trigger large earthquakes and reactivate faults through which CO_2 could leak." They presented four main arguments supporting their case:

- Sedimentary basins are not normally critically stressed;
- The highest injection pressures occur at the start of injection and can be controlled;
- Capillary forces retain CO_2 while allowing water to dissipate;
- CO_2 gradually dissolves in the brine phase.

So, if we limit ourselves to the case of injection into sedimentary basins (weaker rocks with lower coefficients of friction) and consider the fluid dynamics of CO_2-brine systems, then it is more reasonable to argue that large induced earthquakes are not likely to occur. Furthermore, ensuring that pressure increases during injection are carefully controlled and kept below the level that might induce seismic events is a key issue for CO_2 injection projects.

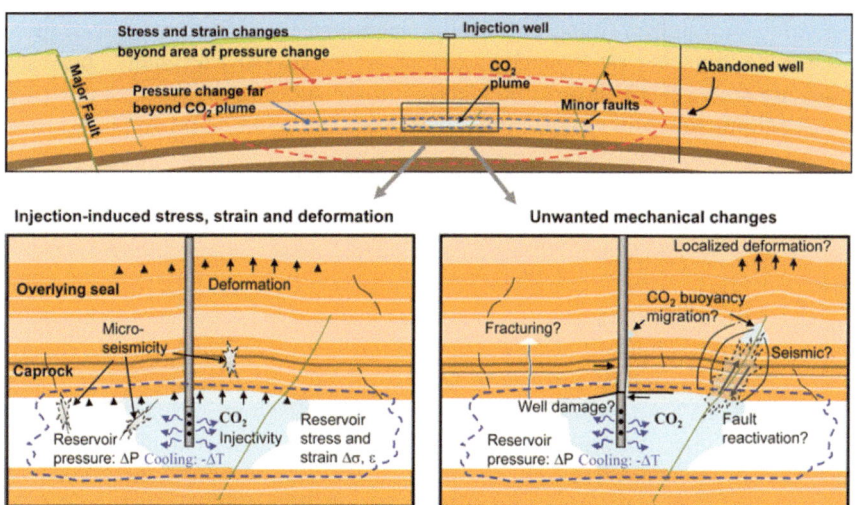

Fig. 2.50 Geomechanical processes and key technical issues associated with CO$_2$ storage in deep sedimentary formations (from Rutqvist 2012). Top: the different regions of influence for a CO$_2$ plume, reservoir pressure changes, and geomechanical changes in a multi-layered system with minor and major faults. Bottom left: injection-induced stress, strain, deformations and potential microseismic events as a result of changes in reservoir pressure and temperature. Bottom right: unwanted inelastic changes that might reduce sequestration efficiency and cause concerns in the local community. Image reproduced with permission © 2019 Springer Nature

Rutqvist (2012) provides an excellent review of the issues involved in understanding and managing the geomechanics of CO$_2$ storage in deep sedimentary formations (Fig. 2.50). Key issues include:

- Understanding the pressure footprint as a function of time,
- Identifying critically aligned fractures and faults near the injection wells.

One of the main uncertainties in understanding the geomechanical response to injection is estimating the in situ stress field and the geomechanical properties of the rock mass well away from the injection well (where measurements are taken). Chiaramonte et al. (2015) provide a very useful analysis of how this geomechanical analysis was done for the Snøhvit injection site in Norway, including the handling of uncertainties in the stress field.

2.7.3   Geomechanical Insights from the in Salah Project

The In Salah CCS Demonstration project in central Algeria was extensively studied as part of a Joint Industry Project from 2007 to 2013. Here we briefly consider the insights gained on the geomechanical aspects of the project. Because of the elevated location of the site (430 m above sea level) in a desert environment with very little vegetation, researchers found that satellite measurements of surface deformation, using the Interferometric Synthetic Aperture Radar (InSAR) technique, provided a very accurate record of pressure changes in the subsurface (Vasco et al. 2008, 2010). Millimetre-scale changes in surface elevation could be related to pressure developments in the injection horizon 2 km below the surface, allowing the nature pressure developments related to CO$_2$ injection to be understood in quite some detail.

The main observations are summarized in Fig. 2.51, showing how surface deformation can be related to rock deformation in the rock mass below. Elevated pressures in the injection unit (between 5 and 7 MPa) led to an expansion of the rock mass producing a detectable deformation on the surface. Most of the deformation was elastic (i.e. recoverable strain) but some deformation was permanent and taken up by displacement of fractures. Ringrose et al. (2013) summarized these insights in a sketch shown in Fig. 2.52. An important observation was that one fracture zone, intersecting the horizontal injection well, was reactivated with some associated microseismicity (Goertz-Allmann et al. 2014). The fracture zone was also evident on the 3D seismic data acquired during the project, revealed as a velocity pull down effect along the

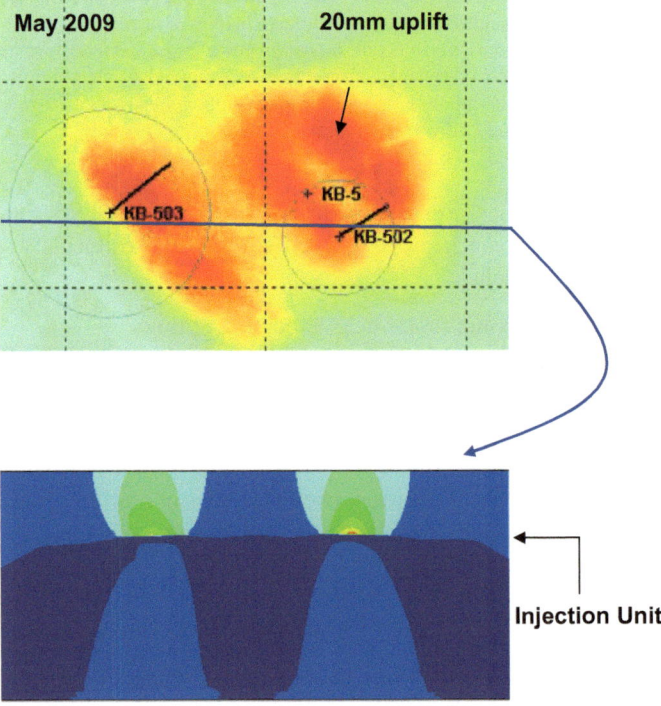

Fig. 2.51 Overview of strain observations at the In Salah CO$_2$ injection site. Upper image: map of InSAR based observations of surface uplift of up to 20 mm in the region over 2 horizontal injection wells (KB-502 and KB-503) Lower image: corresponding rock strain model based on studies by Gemmer et al. (2012). Section is 6 km vertically and 15 km across; green shading indicates expansion (positive strain)

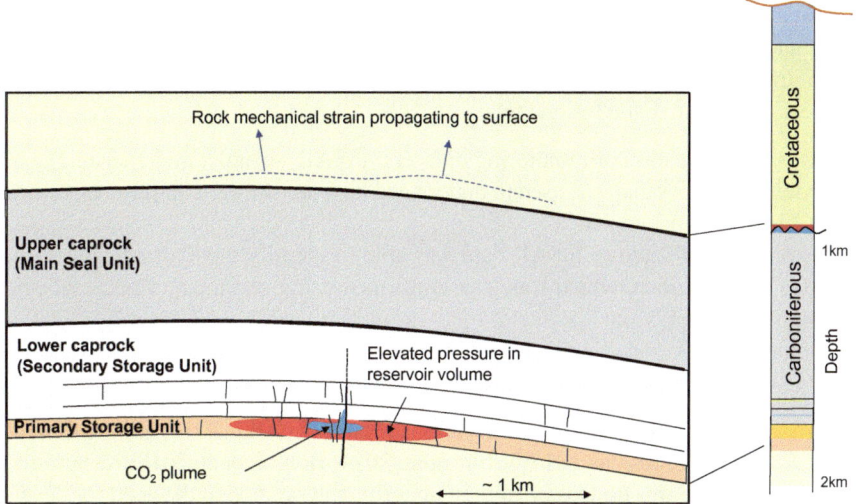

Fig. 2.52 Sketch illustrating the main geomechanical observations around injection well KB-502 at the In Salah injection site

facture zone (Ringrose et al. 2011; White et al. 2014). Numerous modelling studies were conducted (including Vasco et al. 2010; Bissell et al. 2011; Gemmer et al. 2012; Rinaldi and Rutqvist 2013; Bohloli et al. 2017) to better understand the relationships between multiphase flow in the subsurface and the geomechanical response of the rock system (both the rock mass and the fractures). Figure 2.53 summarizes these insights, illustrating how the reference rock mechanical model (rock mass only) was

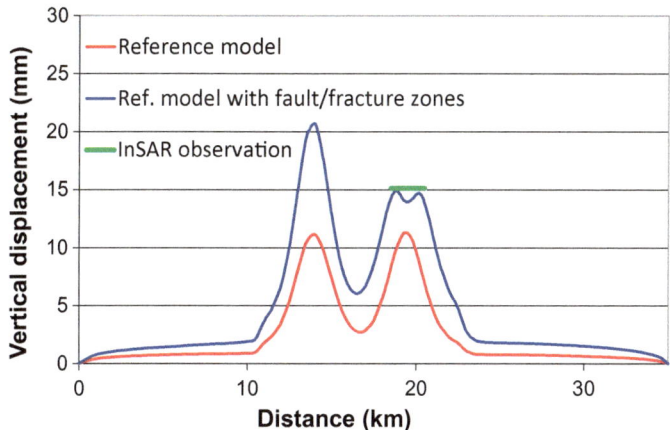

Fig. 2.53 Geomechanical modeling using the In Salah reference model (redrawn from Gemmer et al. 2012) compared with maximum observed uplift from InSAR data (from Rinaldi and Rutqvist 2013)

insufficient to explain the observed deformation, while inclusion of the effects of faults and fracture zones (non-elastic strain) led to much better predictions.

To summarize these insights, especially with regard to potential insights for future projects, we can note the following:

- It was a combination of monitoring datasets (InSAR, seismic, microseismic) and modelling studies that was used to understand how the storage complex responded to CO_2 injection;
- The analysis of satellite InSAR data was able to identify the hydraulically stimulated fracture zone, which was later confirmed using seismic and microseismic monitoring data;
- Geomechanical modelling revealed that the rock mass response was partly elastic strain and partly permanent strain due to deformation along fractures.

Integrated analysis of all the observations (Shi et al. 2012, 2019; Ringrose et al. 2013; Rinaldi and Rutqvist 2013; White et al. 2014; Bohloli et al. 2018) was used to confirm that there was no loss of storage integrity and pointed to the need for developing detailed geological and geomechanical models to understand the processes occurring in the storage unit.

References

Al-Hussainy R, Ramey HJ Jr, Crawford PB (1966) The flow of real gases through porous media. J Pet Technol 624–636

Alnes H, Eiken O, Nooner S, Sasagawa G, Stenvold T, Zumberge M (2011) Results from Sleipner gravity monitoring: updated density and temperature distribution of the CO_2 plume. Energy Procedia 4:5504–5511

Amarasinghe W, Fjelde I., Rydland, JA, Guo Y (2019) Effects of permeability and wettability on CO_2 dissolution and convection at realistic saline reservoir conditions: a visualization study. In: SPE Europec featured at 81st EAGE Conference and Exhibition. Society of Petroleum Engineers

Andersen O, Tangen G, Ringrose P, Greenberg SE (2018) CO_2 data share: a platform for sharing CO2 storage reference datasets from demonstration projects. In: 14th greenhouse gas control technologies conference Melbourne

Arts R, Eiken O, Chadwick A, Zweigel P, Van der Meer L, Zinszner B (2004) Monitoring of CO_2 injected at Sleipner using time-lapse seismic data. Energy 29(9):1383–1392

Bachu S (2015) Review of CO_2 storage efficiency in deep saline aquifers. Int J Greenhouse Gas Control 40:188–202

Bachu S, Bonijoly D, Bradshaw J, Burruss R, Holloway S, Christensen NP, Mathiassen OM (2007) CO_2 storage capacity estimation: methodology and gaps. Int J Greenhouse Gas Control 1(4):430–443

Baines SJ, Worden RH (2004) The long-term fate of CO_2 in the subsurface: natural analogues for CO_2 storage. Geol Soc Lond Spec Pub 233(1):59–85

Bakke S, Øren PE (1997) 3-D pore-scale modelling of sandstones and flow simulations in the pore networks. SPE Journal 2(02):136–149

Baklid A, Korbol R, Owren G (1996) Sleipner vest CO_2 disposal, CO_2 injection into a shallow underground aquifer. In: SPE annual technical conference and exhibition. society of Petroleum Engineers. https://doi.org/10.2118/36600-ms

Baumann G, Henninges J, De Lucia M (2014) Monitoring of saturation changes and salt precipitation during CO_2 injection using pulsed neutron-gamma logging at the Ketzin pilot site. Int J Greenhouse Gas Control 28:134–146

Bennion B, Bachu S (2006) Dependence on temperature, pressure, and salinity of the IFT and relative permeability displacement characteristics of CO_2 injected in deep Saline aquifers. Paper SPE 102138, presented at the 2006 SPE annual technical conference and exhibition, San Antonio, TX, USA, September 24–27

Benson S, Cook P, Anderson J, Bachu S, Nimir HB, Basu B, Bradshaw J, Deguchi G, Gale J, von Goerne G, Heidug W (2005) Underground geological storage. IPCC special report on carbon dioxide capture and storage, pp 195–276

Bentley M (2016) Modelling for comfort? Pet Geosci 22(1):3–10

Bentley M, Ringrose P (2017) Future directions in reservoir modelling: new tools and 'fit-for-purpose' workflows. In: Geological society, London, petroleum geology conference series, vol 8. Geological Society of London, pp PGC8-40

Berg RR (1975) Capillary pressures in stratigraphic traps. AAPG Bull 59(6):939–956

Bickle M, Chadwick A, Huppert HE, Hallworth M, Lyle S (2007) Modelling carbon dioxide accumulation at Sleipner: implications for underground carbon storage. Earth Planet Sci Lett 255(1–2):164–176

Bissell RC, Vasco DW, Atbi M, Hamdani M, Okwelegbe M, Goldwater MH (2011) A full field simulation of the in Salah gas production and CO_2 storage project using a coupled geo-mechanical and thermal fluid flow simulator. Energy Procedia 4:3290–3297

Black JR, Carroll SA, Haese RR (2015) Rates of mineral dissolution under CO_2 storage conditions. Chem Geol 399:134–144

Blunt MJ (2001) Flow in porous media—pore-network models and multiphase flow. Curr Opin Colloid Interface Sci 6(3):197–207

Boait FC, White NJ, Bickle MJ, Chadwick RA, Neufeld JA, Huppert HE (2012) Spatial and temporal evolution of injected CO_2 at the Sleipner Field, North Sea. J Geophys Res Solid Earth 117(B3)

Bohloli B, Ringrose P, Grande L, Nazarian B (2017) Determination of the fracture pressure from CO_2 injection time-series datasets. Int J Greenhouse Gas Control 61:85–93

Bohloli B, Bjørnarå TI, Park J, Rucci A (2018) Can we use surface uplift data for reservoir performance monitoring? A case study from In Salah, Algeria. Int J Greenhouse Gas Control 76:200–207

Bond CE, Wightman R, Ringrose PS (2013) The influence of fracture anisotropy on CO_2 flow. Geophys Res Lett 40(7):1284–1289

Bradshaw J, Bachu S, Bonijoly D, Burruss R, Holloway S, Christensen NP, Mathiassen OM (2007) CO_2 storage capacity estimation: issues and development of standards. Int J Greenhouse Gas Control 1(1):62–68

Busch A, Alles S, Gensterblum Y, Prinz D, Dewhurst DN, Raven MD, Stanjek H, Krooss BM (2008) Carbon dioxide storage potential of shales. Int J Greenhouse Gas Control 2(3):297–308

Carroll SA, McNab WW, Torres SC (2011) Experimental study of cement-sandstone/shale-brine-CO_2 interactions. Geochem Trans 12(1):9

Carruthers D, Ringrose P (1998) Secondary oil migration: oil-rock contact volumes, flow behaviour and rates. Geol Soc Lond Spec Pub 144(1):205–220

Cavanagh A (2013) Benchmark calibration and prediction of the Sleipner CO_2 plume from 2006 to 2012. Energy Procedia 37:3529–3545

Cavanagh AJ, Haszeldine RS (2014) The Sleipner storage site: capillary flow modeling of a layered CO_2 plume requires fractured shale barriers within the Utsira Formation. Int J Greenhouse Gas Control 21:101–112

Cavanagh AJ, Ringrose PS (2011) Simulation of CO_2 distribution at the In Salah storage site using high-resolution field-scale models. Energy Procedia 4:3730–3737

Cavanagh AJ, Haszeldinc RS, Blunt MJ (2010) Open or closed? A discussion of the mistaken assumptions in the Economides pressure analysis of carbon sequestration. J Pet Sci Eng 74(1–2):107–110

Cavanagh AJ, Haszeldine RS, Nazarian B (2015) The Sleipner CO_2 storage site: using a basin model to understand reservoir simulations of plume dynamics. First Break 33(6):61–68

Chadwick RA, Noy DJ (2010) History-matching flow simulations and time-lapse seismic data from the Sleipner CO_2 plume. In: Geological Society, London, petroleum geology conference series, vol 7, issue No 1, pp 1171–1182. Geological Society of London

Chadwick A, Williams G, Delepine N, Clochard V, Labat K, Sturton S, Buddensiek ML, Dillen M, Nickel M, Lima AL, Arts R (2010) Quantitative analysis of time-lapse seismic monitoring data at the Sleipner CO_2 storage operation. Lead Edge 29(2):170–177

Chiaramonte L, White JA, Trainor-Guitton W (2015) Probabilistic geomechanical analysis of compartmentalization at the Snøhvit CO_2 sequestration project. J Geophys Res Solid Earth 120(2):1195–1209

Cooper C, Members of the Carbon Capture Project (2009). A technical basis for carbon dioxide storage: London and New York. Chris Fowler International, pp 3–20. http://www.CO2captureproject.org/

Coward MP, Ries AC (2003) Tectonic development of North African basins. In: Arthur TJ, Macgregor DS, Cameron NR (eds) Petroleum geology of Africa: new themes and developing technologies. Special Publication 207, Geological Society, London, pp 61–83

Durlofsky LJ (1991) Numerical calculation of equivalent grid block permeability tensors for heterogeneous porous media. Water Resour Res 27(5):699–708

EC (2009) Directive 2009/31/EC of the European Parliament and of the Council of 23 April 2009 on the geological storage of carbon dioxide and amending Council Directive 85/337/EEC, European Parliament and Council Directives 2000/60/EC, 2001/80/EC, 2004/35/EC, 2006/12/EC, 2008/1/EC and Regulation (EC) No 1013/2006

Ehlig-Economides C, Economides MJ (2010) Sequestering carbon dioxide in a closed underground volume. J Pet Sci Eng 70(1–2):123–130

Eiken O, Ringrose P, Hermanrud C, Nazarian B, Torp TA, Høier L (2011) Lessons learned from 14 years of CCS operations: Sleipner, In Salah and Snøhvit". Energy Procedia 4:5541–5548

Fjær E, Holt RM, Raaen AM, Risnes R, Horsrud P (2008) Petroleum related rock mechanics, 2nd edn. Elsevier, 514 p

Furre AK, Kiær A, Eiken O (2015) CO_2-induced seismic time shifts at Sleipner. Interpretation 3(3):SS23–SS35. https://doi.org/10.1190/INT-2014-0225.1

Furre AK, Eiken O, Alnes H, Vevatne JN, Kiær AF (2017) 20 years of monitoring CO_2-injection at Sleipner. Energy Procedia 114:3916–3926

Furre A, Ringrose P, Santi AC (2019) Observing the invisible—CO_2 feeder chimneys on seismic time-lapse data. In: 81st EAGE conference and exhibition 2019

Gemmer L, Hansen O, Iding M, Leary S, Ringrose P (2012) Geomechanical response to CO_2 injection at Krechba, In Salah, Algeria. First Break 30(2):79–84

Gibson-Poole CM, Svendsen L, Underschultz J, Watson MN, Ennis-King J, Van Ruth PJ, Nelson EJ, Daniel RF, Cinar Y (2008) Site characterisation of a basin-scale CO_2 geological storage system: Gippsland Basin, southeast Australia. Environ Geol 54(8):1583–1606

Gilfillan SM, Ballentine CJ, Holland G, Blagburn D, Lollar BS, Stevens S, Schoell M, Cassidy M (2008) The noble gas geochemistry of natural CO_2 gas reservoirs from the Colorado Plateau and Rocky Mountain provinces, USA. Geochim Cosmochim Acta 72(4):1174–1198

Goertz-Allmann BP, Kühn D, Oye V, Bohloli B, Aker E (2014) Combining microseismic and geomechanical observations to interpret storage integrity at the In Salah CCS site. Geophys J Int 198(1):447–461

Golan M, Whitson CH (1991) Well performance, 2nd edn. Prentice Hall

Grude S, Landrø M, Dvorkin J (2014) Pressure effects caused by CO_2 injection in the Tubåen Fm., the Snøhvit field. Int J Greenhouse Gas Control 27:178–187

Hansen H, Eiken O, Aasum TA (2005) Tracing the path of carbon dioxide from a gas-condensate reservoir, through an amine plant and back into a subsurface aquifer—case study: the Sleipner area, Norwegian North Sea. Society of Petroleum Engineers, SPE paper 96742. https://doi.org/10.2118/96742-ms

Hansen O, Gilding D, Nazarian B, Osdal B, Ringrose P, Kristoffersen JB, Eiken O, Hansen H (2013) Snøhvit: the history of injecting and storing 1 Mt CO_2 in the Fluvial Tubåen Fm. Energy Procedia 37:3565–3573

Huang Y, Ringrose PS, Sorbie KS (1995) Capillary trapping mechanisms in water-wet laminated rocks. SPE Reservoir Eng 10(4). https://doi.org/10.2118/28942-pa

Huang Y, Ringrose PS, Sorbie KS (1996) The effects of heterogeneity and wettability on oil recovery from laminated sedimentary structures. SPE J 1(04):451–462

Iding M, Ringrose P (2010) Evaluating the impact of fractures on the performance of the In Salah CO_2 storage site. Int J Greenhouse Gas Control 4(2):242–248

Kaszuba JP, Janecky DR, Snow MG (2003) Carbon dioxide reaction processes in a model brine aquifer at 200 C and 200 bars: implications for geologic sequestration of carbon. Appl Geochem 18(7):1065–1080

Kiær AF, Eiken O, Landrø M (2016) Calendar time interpolation of amplitude maps from 4D seismic data. Geophys Prospect 64(2):421–430

Krevor SC, Pini R, Li B, Benson SM (2011) Capillary heterogeneity trapping of CO_2 in a sandstone rock at reservoir conditions. Geophys Res Lett 38(15)

Krevor S, Blunt MJ, Benson SM, Pentland CH, Reynolds C, Al-Menhali A, Niu B (2015) Capillary trapping for geologic carbon dioxide storage—from pore scale physics to field scale implications. Int J Greenhouse Gas Control 40:221–237

Lee WJ, Wattenbarger RA (1996) Gas reservoir engineering, SPE textbook series, vol 5. Society of Petroleum Engineers, 349

Lopez O, Idowu N, Mock A, Rueslåtten H, Boassen T, Leary S, Ringrose P (2011) Pore-scale modelling of CO 2-brine flow properties at In Salah, Algeria. Energy Procedia 4:3762–3769

Marcussen Ø, Faleide JI, Jahren J, Bjørlykke K (2010) Mudstone compaction curves in basin modelling: a study of mesozoic and cenozoic sediments in the northern North Sea. Basin Res 22(3):324–340

Mathieson A, Midgley J, Dodds K, Wright I, Ringrose P, Saoul N (2010) CO_2 sequestration monitoring and verification technologies applied at Krechba, Algeria. Leadv Edge 29(2):216–222

Meckel TA, Bryant SL, Ganesh PR (2015) Characterization and prediction of CO_2 saturation resulting from modeling buoyant fluid migration in 2D heterogeneous geologic fabrics. Int J Greenhouse Gas Control 34:85–96

Metz B (ed) (2005) Carbon dioxide capture and storage: special report of the intergovernmental panel on climate change. Cambridge University Press

Miri R, van Noort R, Aagaard P, Hellevang H (2015) New insights on the physics of salt precipitation during injection of CO_2 into saline aquifers. Int J Greenhouse Gas Control 43:10–21

Naylor M, Wilkinson M, Haszeldine RS (2011) Calculation of CO_2 column heights in depleted gas fields from known pre-production gas column heights. Mar Pet Geol 28(5):1083–1093

Nazarian B, Held R, Høier L, Ringrose P (2013) Reservoir management of CO_2 injection: pressure control and capacity enhancement. Energy Procedia 37:4533–4543

Niemi A, Bear J, Bensabat J (2017) Geological storage of CO_2 in Deep Saline Formations. Springer

Nordbotten JM, Celia MA (2006) Similarity solutions for fluid injection into confined aquifers. J Fluid Mech 561:307–327

Nordbotten JM, Celia MA (2012) Geological storage of CO_2: modeling approaches for large-scale simulation. Wiley

Nordbotten JM, Celia MA, Bachu S (2005) Injection and storage of CO_2 in deep saline aquifers: analytical solution for CO_2 plume evolution during injection. Transp Porous Media 58(3):339–360

Okwen RT, Stewart MT, Cunningham JA (2010) Analytical solution for estimating storage efficiency of geologic sequestration of CO_2. Int J Greenhouse Gas Control 4(1):102–107

Oldenburg CM, Mukhopadhyay S, Cihan A (2016) On the use of Darcy's law and invasion-percolation approaches for modeling large-scale geologic carbon sequestration. Greenhouse Gases Sci Technol 6(1):19–33

Pau GS, Bell JB, Pruess K, Almgren AS, Lijewski MJ, Zhang K (2010) High-resolution simulation and characterization of density-driven flow in CO_2 storage in saline aquifers. Adv Water Resour 33(4):443–455

Pawar RJ, Bromhal GS, Carey JW, Foxall W, Korre A, Ringrose PS, Tucker O, Watson MN, Mathieson A, White JA (2015) Recent advances in risk assessment and risk management of geologic CO_2 storage. Int J Greenhouse Gas Control 40:292–311

Pickup GE, Sorbie KS (1996) The scaleup of two-phase flow in porous media using phase permeability tensors. SPE J 1(04):369–382

Pickup GE, Ringrose PS, Jensen JL, Sorbie KS (1994) Permeability tensors for sedimentary structures. Math Geol 26(2):227–250

Rapoport LA (1955) Scaling laws for use in design and operation of water-oil flow models. Pet Trans 145–150

Reynolds CA, Krevor S (2015) Characterizing flow behavior for gas injection: relative permeability of CO-brine and N_2-water in heterogeneous rocks. Water Resour Res 51(12):9464–9489

Riaz A, Hesse M, Tchelepi HA, Orr FM (2006) Onset of convection in a gravitationally unstable diffusive boundary layer in porous media. J Fluid Mech 548:87–111

Rinaldi AP, Rutqvist J (2013) Modeling of deep fracture zone opening and transient ground surface uplift at KB-502 CO_2 injection well, In Salah, Algeria. Int J Greenhouse Gas Control 12:155–167

Ringrose PS (2018) The CCS hub in Norway: some insights from 22 years of saline aquifer storage. Energy Procedia 146:166–172

Ringrose P, Bentley M (2016). Reservoir model design. Springer

Ringrose PS, Sorbie KS, Corbett PWM, Jensen JL (1993) Immiscible flow behaviour in laminated and cross-bedded sandstones. J Petrol Sci Eng 9(2):103–124

Ringrose PS, Yardley G, Vik E, Shea WT, Carruthers DJ (2000) Evaluation and benchmarking of petroleum trap fill and spill models. J Geochem Explor 69–70:689–693. https://doi.org/10.1016/S0375-6742(00)00072-8

Ringrose P, Atbi M, Mason D, Espinassous M, Myhrer Ø, Iding M, Wright I (2009) Plume development around well KB-502 at the In Salah CO_2 storage site. First Break 27(1):85–89

Ringrose PS, Roberts DM, Gibson-Poole CM, Bond C, Wightman R, Taylor M, Østmo S (2011) Characterisation of the Krechba CO_2 storage site: critical elements controlling injection performance. Energy Procedia 4:4672–4679

Ringrose PS, Mathieson AS, Wright IW, Selama F, Hansen O, Bissell R, Saoula N, Midgley J (2013) The In Salah CO_2 storage project: lessons learned and knowledge transfer. Energy Procedia 37:6226–6236

Ringrose P, Greenberg S, Whittaker S, Nazarian B, Oye V (2017) Building confidence in CO_2 storage using reference datasets from demonstration projects. Energy Procedia 114:3547–3557

Rutqvist J (2012) The geomechanics of CO_2 storage in deep sedimentary formations. Geotech Geol Eng 30(3):525–551

Sahasrabudhe SN, Rodriguez-Martinez V, O'Meara M, Farkas BE (2017) Density, viscosity, and surface tension of five vegetable oils at elevated temperatures: measurement and modeling. Int J Food Prop 20(sup2):1965–1981

Sclater JG, Christie P (1980) Continental stretching: an explanation of the post-mid-cretaceous subsidence of the central North Sea Basin. J Geophys Res Solid Earth 85(B7):3711–3739 (1978–2012)

Shi JQ, Sinayuc C, Durucan S, Korre A (2012) Assessment of carbon dioxide plume behaviour within the storage reservoir and the lower caprock around the KB-502 injection well at In Salah. Int J Greenhouse Gas Control 7:115–126

Shi JQ, Durucan S, Korre A, Ringrose P, Mathieson A (2019) History matching and pressure analysis with stress-dependent permeability using the In Salah CO_2 storage case study. Int J Greenhouse Gas Control 91:102844

Shook M, Li D, Lake LW (1992) Scaling immiscible flow through permeable media by inspectional analysis. In Situ 16(4):311–311

Singh VP, Cavanagh A, Hansen H, Nazarian B, Iding M, Ringrose PS (2010) Reservoir modeling of CO_2 plume behavior calibrated against monitoring data from Sleipner, Norway. In: SPE annual technical conference and exhibition. Society of Petroleum Engineers. https://doi.org/10.2118/134891-MS

Stephen KD, Pickup GE, Sorbie KS (2001) The local analysis of changing force balances in immiscible incompressible two-phase flow. Transp Porous Media 45(1):63–88

Torp TA, Gale J (2004) Demonstrating storage of CO_2 in geological reservoirs: the Sleipner and SACS projects. Energy 29(9–10):1361–1369

Trevisan L, Pini R, Cihan A, Birkholzer JT, Zhou Q, Illangasekare TH (2015) Experimental analysis of spatial correlation effects on capillary trapping of supercritical CO_2 at the intermediate laboratory scale in heterogeneous porous media. Water Resour Res 51(11):8791–8805

Van der Meer LGH (1995) The CO_2 storage efficiency of aquifers. Energy Convers Manag 36(6):513–518

Vasco DW, Ferretti A, Novali F (2008) Reservoir monitoring and characterization using satellite geodetic data: interferometric synthetic aperture radar observations from the Krechba field, Algeria. Geophysics 73(6):WA113–WA122

Vasco DW, Rucci A, Ferretti A, Novali F, Bissell RC, Ringrose PS, Mathieson AS, Wright IW (2010) Satellite-based measurements of surface deformation reveal fluid flow associated with the geological storage of carbon dioxide. Geophys Res Lett 37(3)

Verdon JP, Kendall JM, Stork AL, Chadwick RA, White DJ, Bissell RC (2013) Comparison of geomechanical deformation induced by megatonne-scale CO_2 storage at Sleipner, Weyburn, and In Salah. Proc Natl Acad Sci 110(30):E2762–E2771

Vilarrasa V, Carrera J (2015) Geologic carbon storage is unlikely to trigger large earthquakes and reactivate faults through which CO_2 could leak. Proc Natl Acad Sci 112(19):5938–5943

Vilarrasa V, Bolster D, Dentz M, Olivella S, Carrera J (2010) Effects of CO_2 compressibility on CO_2 storage in deep saline aquifers. Transp Porous Media 85(2):619–639

White JA, Chiaramonte L, Ezzedine S, Foxall W, Hao Y, Ramirez A, McNab W (2014) Geomechanical behavior of the reservoir and caprock system at the In Salah CO2 storage project. Proc Natl Acad Sci 111(24):8747–8752

Wilkinson D, Willemsen JF (1983) Invasion percolation: a new form of percolation theory. J Phys A: Math Gen 16(14):3365

Wilkinson M, Haszeldine RS, Fallick AE, Odling N, Stoker SJ, Gatliff RW (2009) CO_2–mineral reaction in a natural analogue for CO_2 storage—implications for modeling. J Sediment Res 79(7):486–494

Williams GA, Chadwick RA (2017) An improved history-match for layer spreading within the Sleipner plume including thermal propagation effects. Energy Procedia 114:2856–2870

Yortsos YC (1995) A theoretical analysis of vertical flow equilibrium. Transp Porous Media 18(2):107–129

Zhou D, Fayers FJ, Orr FM Jr (1997) Scaling of multiphase flow in simple heterogeneous porous media. SPE Reservoir Eng 12(03):173–178

Zhou Q, Birkholzer JT, Tsang CF, Rutqvist J (2008) A method for quick assessment of CO_2 storage capacity in closed and semi-closed saline formations. Int J Greenhouse Gas Control 2(4):626–639

Zoback MD (2007) Reservoir geomechanics. Cambridge University Press, Cambridge, 449 p

Zoback MD, Gorelick SM (2012) Earthquake triggering and large-scale geologic storage of carbon dioxide. Proc Natl Acad Sci 109(26):10164–10168

Zweigel P, Arts R, Lothe AE, Lindeberg EB (2004) Reservoir geology of the Utsira Formation at the first industrial-scale underground CO_2 storage site (Sleipner area, North Sea). Geol Soc Lond Spec Publ 233(1):165–180

Chapter 3
CO$_2$ Storage Project Design

Designing real CO$_2$ injection projects and maturing them from the concept stage to the execution stage is a major undertaking, but at the same time based on well-known practice. During the 'petroleum age' (Fig. 1.1) human beings developed over 60,000 oil and gas fields using several million wells—so there is plenty of established practice in well and reservoir engineering (see for example Golan and Whitson 1991 or Dake 2001). What we aim to do in this short introduction is identify some important aspects of CO$_2$ storage project design based on insights from 'early-mover' projects. Hopefully, in the future, more complete treatments of this topic will emerge as the currently rather small CO$_2$ disposal industry develops into a globally significant activity.

3.1 Injection Well Design

Let's start with the assumption that we need to design, drill and operate one or more injection wells to meet a given injection target. We assume the general principles developed for petroleum industry well design and focus on the key issues for CO$_2$ storage, namely:

- Well design
- Well placement
- Injectivity
- Well integrity.

The overarching question is—How do you design an injection well that ensures:

- Safe operation
- Sustained capacity
- Operational reliability?

P. Ringrose, *How to Store CO$_2$ Underground: Insights from early-mover CCS Projects*, SpringerBriefs in Earth Sciences, https://doi.org/10.1007/978-3-030-33113-9_3

Although the oil and gas industry has experience from millions of production wells and many 100,000's of water/gas injection wells, CO$_2$ injection experience is more limited. There are probably of order 10,000 wells used in CO$_2$ EOR projects, mainly in USA and Canada, but only a few 10's of CO$_2$ injection wells designed for dedicated storage. Some of this operational experience from CO$_2$ EOR projects has been reviewed in the public domain (Bachu and Watson 2009) and Michael et al. (2010) provide a useful overview of CO$_2$ injection at pilot and commercial operations. Here we will only consider wells used for CO$_2$ injection into saline aquifer formations, acknowledging that experience from CO$_2$ EOR projects is also very relevant.

To encapsulate the key topics that have merged from the early experience with CO$_2$ injection wells, we can identify four main themes:

- Reservoir heterogeneity effects (e.g. lower than expected permeability or the presences of reservoir flow barriers);
- Rock mechanical effects (determining a safe down-hole injection pressure and understanding how the rock system respond to elevated injection pressures);
- CO$_2$-brine reactions (especially the potential for corrosion and salt precipitation);
- Thermal effects (what happens if the CO$_2$ is significantly hotter or colder than reservoir).

A good place to start is to review the design choices for the Sleipner CO$_2$ injection well 15/9-A16—the world's first commercial offshore CO$_2$ injection well (Fig. 3.1). Key elements of the well design (Hansen et al. 2005) were:

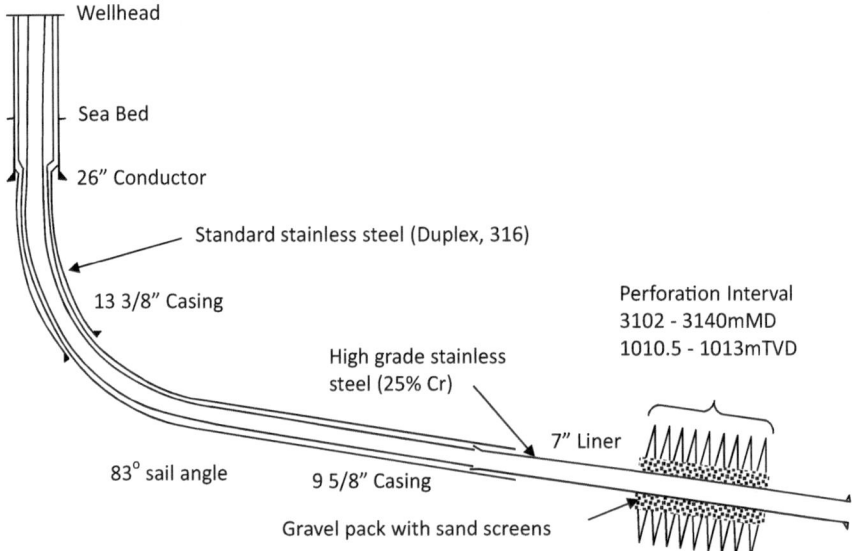

Fig. 3.1 Summary of the Sleipner CO$_2$ injection well 15/9-A16 design (redrawn from Hansen et al. 2005)

- Use of a long-reach horizontal well (with a sail angle of 83°) to ensure storage well away from the gas production platform and beneath the intended structural closure.
- Perforated injection interval of 38 m with the top of the injection interval at a depth of 1010 m, somewhat shorter than the original design plan for a 100 m completion interval.
- Use of standard stainless steel (Duplex-grade) for most of the well components, but selection of high-grade (25% Cr) steel for the $7''$ injection tubing and the exposed sections of the $9\,^5/_8''$ well casing, in order to ensure a good level of corrosion resistance.

This well has performed very reliably for over 20 years, despite the early challenges related to injectivity (reviewed in Sect. 2.6 above), attesting to the value of careful choice of steel quality to ensure long-term relatability.

Subsequent projects (e.g. Snøhvit, In Salah, Quest) have developed similar designs, but with a range of choices on well placement—near-vertical or deviated wells at Snøhvit and Quest and horizontal wells at the In Salah project.

The choice of well path is a mix of the need to access suitable geology and to ensure sufficient injectivity. The injectivity of a well (Sect. 2.6) is proportional to the formation permeability and the thickness of the injection interval (Eq. 2.34)—often referred to as the 'k-h product' or just 'kh'. For lower permeability formations you need a longer injection interval to achieve an acceptable kh.

Long-reach horizontal wells were chosen at the In Salah project (Wright et al. 2009; Ringrose et al. 2013) to achieve sufficient injectivity in a low-permeability sandstone (<10 md) and to intersect the natural (stress-aligned) fracture direction. A vertical well was preferred for the first injection well at the Snøhvit site (Hansen et al. 2013) in order to intersect multiple sand intervals in the same well. So, the well placement choice is very much site dependent. However, a major difference for CO_2 injection wells as compared with oil and gas wells is the depth of the injection interval. Oil and gas production wells typically target the highest parts of the reservoir interval in order benefit from fluid buoyancy forces, while water injectors may be placed down dip. For CO_2 injection projects, it is now clear that injection in deeper units is preferable, allowing CO_2 to slowly migrate upwards into the geological layers—to some extent replicating oilfield filling processes over geological time. This was the design choice at Sleipner where injection at the base of the Utsira sandstone unit allowed a stack of sandstone units to be exploited by a single well (see Sect. 2.5.4 above). Figure 3.2 summarizes these well-placement options—essentially to target the good quality sandstone (or carbonate) units and then choose a placement and well angle to optimize injectivity and enable upwards migration of CO_2 into the brine filled aquifer.

It is important not to forget the well-cement components of the well design—casing needs to be secured to the formation using appropriate cements and good cement placement practice. This aspect of well design will be discussed below as part of well integrity (Sect. 3.3). So, to summarize the well-design issues: on the whole, standard petroleum-industry well-design best practice can be used, but there

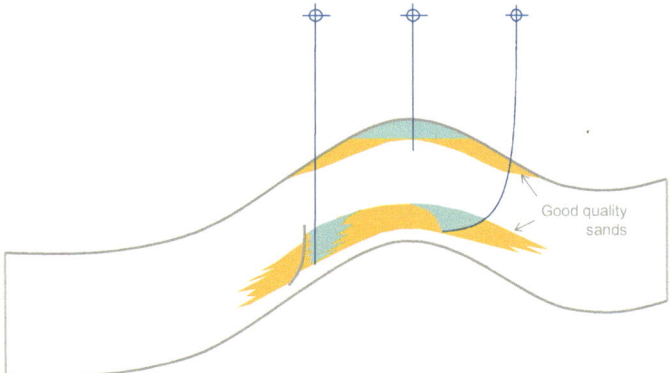

Fig. 3.2 Simple sketch of injection well placement options

are some important differences that lead to CO_2-specific well designs, which can be summarized as follows:

1. Well placement chosen to maximize both injectivity and utilisation of the storage unit—typically by injection towards the base of permeable units;
2. Material choices for well components that have a higher resistance to corrosion than many standard wells, using stainless steel or high chromium steels for exposed sections of tubing or casing;
3. Overall well designs to ensure secure, long-term containment, using good choices for cement placement.

3.2 CO$_2$ Thermodynamics and Transport

Understanding the phase behaviour of CO_2 is fundamental to CO_2 storage operations. The whole concept behind CCS is to store (or dispose of) CO_2 generated from human activities, where the CO_2 is in the gas phase at close to atmospheric conditions, and to store it deep underground where it will be in the liquid or dense phase. This phase transition leads to very effective use of the pore space as CO_2 in the subsurface occupies a much smaller volume than at the surface (refer to Fig. 2.1). However, CO_2 in its liquid or dense form is not a substance we are familiar with in everyday life. As a liquid CO_2 is colourless (looks a bit like water at surface conditions) but is less dense than water (a bit like oil on water) but then much less viscous than oil or water such that flows a bit like a gas. Not a substance we are so familiar with at the surface, but nevertheless a naturally-occurring substance in the subsurface (there are many naturally-occurring accumulations of liquid/dense-phase CO_2 on our planet!).

Therefore, to handle CO_2 in real transport and injection projects we then need to appreciate the CO_2 phase diagram (Fig. 3.3). After the capture process (see Sect. 1.5 above) CO_2 needs to be compressed for transportation to the injection site. At the Snøhvit site, for example, CO_2 is compressed to over 80 bars at the onshore processing facility and is then transported along a 150 km pipeline, as a liquid phase, entering the wellhead at around 140 bars, still in the liquid phase. At bottom-hole conditions (with the pressure around 350 bars at a depth of 2400 m) the CO_2 then flows into the reservoir, moving into the dense phase as the temperatures rises above the critical point (this phase is also called super-critical CO_2). With an initial reservoir temperature of 95 °C the CO_2 warms by around 50 degrees as it enters the reservoir unit. See Maldal and Tappel (2004) and Hansen et al. (2013) for a fuller description of this project.

At the Sleipner CCS project, the wellhead pressure is much lower, at around 62 bars, because the injection formation is much shallower (1000 m deep). With a surface temperature of around 25 °C at the gas processing facility on the Sleipner platform, the CO_2 is initially close to the vapour line at the wellhead, with two-phase flow conditions occurring at this point (Eiken et al. 2011; Lindeberg 2011). With increasing temperature and pressure in the well, the CO_2 soon enters the dense phase and stays in the dense phase in the reservoir. The bottom-hole temperature is estimated (not measured) at about 48 °C, but then the CO_2 cools towards around 35 °C as it within the sandstone reservoir. The behaviour of CO_2 in the multi-layer reservoir at Sleipner is quite complex and strongly dependent on the temperature and

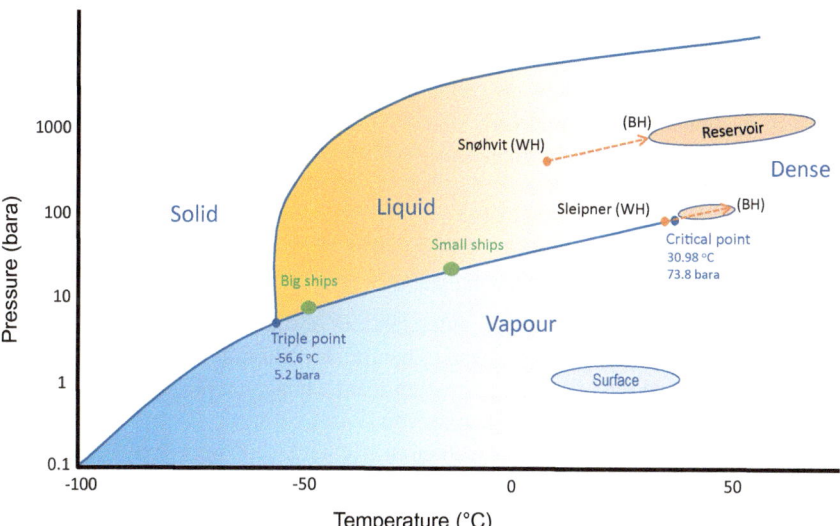

Fig. 3.3 CO_2 phase diagram with typical paths and conditions for CO_2 storage operations. Approximate wellhead (WH) and bottom-hole (BH) conditions for the Sleipner and Snøhvit projects are indicated

flow-property assumptions (for a fuller analysis see, for example, Boait et al. 2012; Singh et al. 2010; Williams and Chadwick 2017).

In the case of transportation of CO_2 by ship (or by road or rail-freight tanker), the CO_2 is compressed to be held just above the boiling line (Fig. 3.3), with smaller shipping solutions opting for higher pressure and higher temperature (e.g. around −20 °C and 20 bar for ships of around 1000–2000 tonne). But for larger-scale shipping with larger tanks the pressure needs to be reduced, requiring chilling the CO_2 to around −45 °C. At the storage site the CO_2 will then need further compression to reach the required injection wellhead pressure. These transport and surface handling solutions for CO_2 storage operations all use established technology, but clearly a good appreciation of phase behaviour is essential. Because real CO_2 capture streams are not pure and have minor impurities (e.g. CH_4, H_2S and H_2O), the accurate determination of the phase behaviour becomes more challenging (see De Visser et al. 2008; Chapoy et al 2013).

When it comes to making detailed engineering assessments of CO_2 transport and storage systems, we need an equation of state (EOS) to make specific assessments or to continuously track the phase behaviour. An EOS is a reference function describing the properties of a substance, or mixture, based on thermodynamical theory and experimental data. There are several options for a CO_2 equation of state:

- Peng-Robinson and Soave-Redlich-Kwong are two commonly-used cubic equations of state, which because they are relatively simple to implement are widely used in modelling packages (see Li and Yan 2009, for a discussion of different cubic EOS equations used for handling CCS processes)
- Span and Wagner (1996) proposed a more accurate but also more demanding EOS for CO_2 which is often used in assessing detailed system behaviour and complex mixtures (Span et al. 2013).

When it comes to designing CO_2 handling systems, especially compression of CO_2 it is important to use the pressure-enthalpy version of the CO_2-phase diagram—something we will not cover here, except to say that compression is usually the most energy demanding component of a CO_2 storage project.

It may be useful to conclude this short section on CO_2 thermodynamics by noting the main lessons learned from early mover CCS projects:

- Firstly, all projects require a good understanding of the phase behaviour, not only for the overall system design but also to handle operational changes—rate fluctuations and thermal effects can lead to significant changes in properties of the CO_2 stream (both in the well and in the reservoir).
- Reducing the water content in the CO_2 stream (e.g. to <100 ppm) may be desirable to reduce corrosion effects but can then lead to salt precipitation in the reservoir, as experienced at the Snøhvit project (Hansen et al. 2013). Often a compromise is needed regarding the degree of drying that is actually necessary.
- The effects of minor components in the CO_2-rich stream can have a significant impact on the phase behavior. For example, a few percent of CH_4 in the injection stream (as occurs at Sleipner; Eiken et al. 2011) can change the critical point or boiling line by several degrees (°C).

- Many operational issues concern how to handle wells and pipelines during planned or unplanned stoppages or in the case of unexpected leakages.

However, despite these operational challenges, extensive guidelines on safe practice exist for operation of CO$_2$ transport and handling systems (e.g. Eldevik et al 2009 or Johnsen et al 2011) based on over 50 years of operational experience in handing of CO$_2$ streams in pipelines and wells.

3.3   Framework for Managing the Storage Site

We have been developing the argument that CO$_2$ storage is a mature and proven technology, and also a highly desirable activity from the point of view of protecting the Earth's atmosphere and mitigating climate change. When proposing or planning a CO$_2$ storage project, discussion soon moves towards the 'what if?' questions:

- What if there is a leak from the storage site?
- What if the plume moves faster than expected and out of the intended storage unit?
- Could the pressure rise above the fracture pressure and cause leakage along a fracture?
- Could earthquakes be provoked by the injection?

These are not easy questions to answer. However, it is important to approach the problem from a sound basis. Wild speculation is not very helpful, but careful analysis of processes and risks should allow each project to mature to practical decisions. In the end a real storage site has be licenced and permitted—that is to say, it must operate within the law. Below we will review the legal and physical aspects of managing a CO$_2$ storage site. But before we do that it is helpful to remind ourselves of a really basic issue—understanding risk. Let's assume there is some level risk associated with operating a CO$_2$ storage site, then the question is 'Is that level of risk acceptable?' There are several ways we can assess the level of acceptable risk:

1. *The climate protection argument*: 'Putting CO$_2$ in deep geological formations is a lot safer and better than putting the same CO$_2$ into the atmosphere.' This is clearly the fundamental argument—but if the climate protection argument is not supported by legal and financial arguments it may not achieve very much on its own.
2. *The legal argument*: If the CO$_2$ storage project complies with the regulatory agreements in place and as long as 'no significant irregularity' occurs, then the risks are acceptable. We are used to permitting other activities, like road safety or oilfield operations, and so CO$_2$ storage can also become an activity which can operate within the law and with a socially-agreed safety level.
3. *The economic argument*: Assuming there is a cost for emitting CO$_2$ to atmosphere (a carbon price) there is then a financial benefit for not emitting that CO$_2$ and consequently CO$_2$ storage has a value. The more an activity has value, the more likely that the associated risks with doing that activity will be accepted.

In practice, all three arguments must be in place for a CO$_2$ storage project to proceed successfully. With no benefits in place no level of risk is likely to be accepted. The CO$_2$ storage projects that have succeeded so far all occur in national jurisdictions where some kind of legal and financial framework is in place (notably, Norway, Canada, USA and Australia).

To give one important example of a legal framework for storage, the *European Directive on the geological storage of carbon dioxide* (EC 2009) adopts the following concepts regarding the 'Storage Complex' meaning the geological system relevant to containment:

- *Sufficient data* shall be accumulated to construct a volumetric and three-dimensional static (3-D)-earth model for the storage site and storage complex, including the caprock, and the surrounding area, including the hydraulically connected areas.
- *Leakage* is defined as 'any release of CO$_2$ from the storage complex', while *significant irregularity* means 'any irregularity in the injection or storage operations or in the condition of the storage complex itself, which implies the risk of a leakage or risk to the environment or human health'.

Although frequently referred to in the directive, the definition of the *Storage Complex* is left rather open—and intended to be defined for each specific case. Figure 3.4 illustrates what is meant by the storage complex—a geological domain including storage units, sealing units and structural closures relevant for demonstrating conformance and containment. So, put simply, the legal framework is that no *significant irregularity* which implies the risk of a leakage or risk to the environment or human health should occur with reference to leakage out of the *storage complex*.

In assessing those risks in detail there are many activities involved, including:

1. Adequate characterisation of the site;
2. Modelling or assessment of the likely future development of the CO$_2$ stored at the site;
3. A method to quantify and assess the risks involved;
4. A decision to proceed with an acceptable level of risk.

Pawar et al. (2015) provide a useful review of risk assessment and risk management of geological CO$_2$ storage, arguing that risks associated with a properly characterized and permitted storage system are extremely low and that sufficient project experience and risk assessment procedures are in place for maturing projects globally. More often it is not technical risks that have led to project failure—the dominant risks are market failure risks and lack of effective communication. That is to say, the real challenge is to develop and communicate the benefits of CO$_2$ storage—the technical risks are actually small and manageable.

Nevertheless, to address the 'what if?' questions about safe operation and management of CO$_2$ storage sites, it is very helpful establish a good understanding of the physical processes involved and to explain what has be learned from early mover projects.

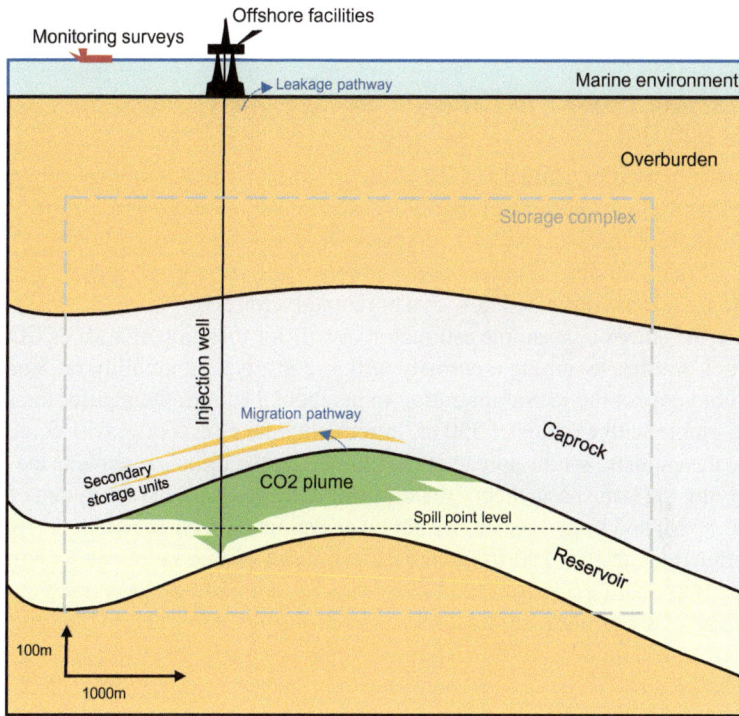

Fig. 3.4 Illustration of what is meant by the storage complex (for an offshore storage setting). Leakage concerns CO_2 flux out of the storage complex, while migration refers to CO_2 fluxes within the storage complex

A major difference between CO_2 injection projects and gas or oil production projects, is that for CO_2 storage there is generally much less well control (e.g. 1 or 2 injection wells) and yet there needs to be some level of confidence about the CO_2 remaining within the storage domain (well away from the wells). In some of the early research pilot projects, such as Otway in Australia (Sharma et al. 2011; Jenkins et al, 2017) and Ketzin in Germany (Ivanova et al. 2012; Martens et al. 2014), dedicated monitoring wells were drilled to check how the CO_2 behaved in the subsurface. However, in general and for large-scale commercial projects we will need to minimize the drilling of observation wells and mainly rely on remote detection and modelling approaches. That is to say, a combination of fluid flow modelling and geophysical/geochemical monitoring will need to be sufficient to have confidence about the site. In the next section we will focus on monitoring approaches, and here we will briefly cover the practices needed to model and understand the CO_2 plume and associated pressure footprint.

3.4   Approaches to Forecasting Site Behaviour

3.4.1   Using Physics to Estimate Storage Site Dimensions

To obtain a first order estimate of the likely footprint of a CO$_2$ storage project, the idealised analytical solutions proposed by Nordbotten et al. (2005) can be used to obtain a quick estimate of the likely spread of a plume. Using Eq. 2.23 (Sect. 2.4.4) we can estimate the lateral extent of the plume, r_{max}, for a given injection volume (assuming viscous-dominated flow in a horizontal aquifer).

Figure 3.5 shows an example estimation of r_{max} for injection of 1 Mt of CO$_2$ into 50 m thick aquifer, assuming a porosity of 0.3. For a typical mobility ratio of $\lambda_r = 4$ we would expect the lateral migration to be about 11 times the aquifer thickness, giving a plume with a radius of 550 m. Introducing the effects of gravity forces will increase the estimate for the spread of the plume, but the principle remains the same.

Applying the same basic theory to the plume at Sleipner for the time point of 2008 when 10.56 Mt had been injected, we obtain an estimate for rmax of around 18 times the aquifer thickness of 190 m, giving an estimated plume radius of 3.4 km. The observed maximum extent of the plume from seismic data at 2008 (ref Fig. 2.22) was around 2.8 km northward extension from the injection point. This indicates the theoretical estimate is 'of the right order of magnitude'—but we must emphasize that the real plume is complex and multi-layered and controlled by vertical migration points and local structural closures (Furre et al. 2019).

Bickle et al. (2007) also proposed an analytical solution for a CO$_2$ plume for the case of gravity-driven flows, noting that for a constant injection rate the radius of the CO$_2$ layer is proportional to the square root of time. Williams et al. (2018) presented a comparison between this analytical solution and various flow simulation tools (comparing black-oil and fully compositional models) for the case of the Sleipner plume, revealing an overall good match between analytical models, flow simulation methods and seismic observations.

Fig. 3.5   Extent of plume for 1Mt injection into a 50 m thick aquifer

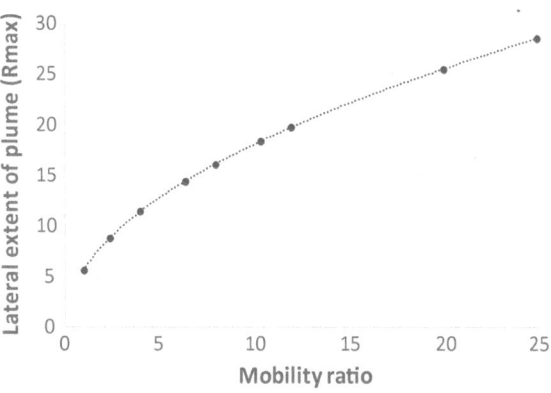

Thus, analytical approaches are useful in giving a likely dimension to the expected plume footprint, but detailed assessment of this problem requires more realistic geological models and flow simulation methods that take into account the complex interactions between the fluid properties (in situ phase behaviour) and the reservoir architecture (flow units with variable petrophysical properties).

3.4.2 Using Flow Simulation to Forecast Scenarios

For geologically realistic cases and for site-specific project designs, we will generally need to use flow simulation approaches to assess the likely growth of the plume. Multiphase flow simulation for oil and gas field developments is a mature technology, covered in numerous textbooks (e.g. Peaceman 2000; Fanchi 2005), and the adaptation of these approaches for simulation of CO_2 storage is also now fairly mature (e.g. Pruess et al. 2004) and discussed above in Sect. 2.5. Here, we will focus on important insights gained on how to apply reservoir simulation methods as part of the CO_2 storage project design.

We can identify some important differences between conventional practice for reservoir flow simulation for hydrocarbon field development compared with the CO_2 storage problem. Key flow simulation issues for CO_2 storage projects are as follows:

1. CO_2 injection projects generally start with very few wells (perhaps none initially) and generally need to forecast expected behaviour out from the injection well and with very few well-control constraints. This contrasts with oilfield modelling is often focused on history matching observations from multiple wells.
2. Injection of a low viscosity fluid (CO_2) into a higher viscosity fluid (brine) gives an unstable mobility ratio and inherently leads to fingering behaviour which is hard to predict. It is actually rather fortunate that this 'initially unstable flow regime' quickly stabilises due to gravity segregation, but nevertheless the flow process is difficult to predict precisely.
3. The properties of the CO_2 phase in the storage unit are very dependent on the in situ pressure and temperature, especially for shallower injection close to the critical point (as with the Sleipner case, Sect. 3.2 above).

This setting makes forecasting CO_2 storage behaviour somewhat more challenging than oil field simulation—we generally have less subsurface information available and yet we need to forecast an unstable displacement of a fluid with uncertain fluid properties. Despite this overall challenge, early projects have found that forecasting CO_2 storage behaviour is possible as long as the uncertainties are appreciated. Another way to put this is that flow simulation for CO_2 storage projects is inherently a forecasting activity, where the objectives are to understand the 'expected behaviour' and the likely 'range of outcomes'.

In their review of three large-scale storage projects, Eiken et al. (2011) pointed out that the actual CO_2 plume development at the Sleipner, In Salah and Snøhvit projects was strongly controlled by unpredicted geological factors that were learned about during injection from monitoring data (especially seismic data). This insight changes the perception that people may initially have about the logic of CO_2 storage, namely to (a) first adequately characterise the site and then (b) monitor just to check storage proceeds 'according to plan.' Project experience so far clearly points to the need to be much more interactive about the project design, such that site monitoring data should be used to learn and update the reservoir model as well as being used to monitor fluid saturation and pressure.

This more interactive way of dynamic reservoir modelling with continuous updates and feedback to the static rock model is actually consistent with the way subsurface modelling is evolving across all disciplines (hydrocarbon fields, water resources and CO_2 storage). In explaining this trend, Bentley and Ringrose (2017) argue for the use of nimble and interactive workflows which are decision led and which allow frequent updating.

Flow simulation for forecasting CO_2 storage therefore needs to have a realistic framework for achieving specific decision-led objectives. The key objectives and questions to address are usually:

- *To understand the overall footprint of the project*: How far is the plume likely to extend? Here analytical approaches can give a first approximation and flow simulations should be focussed on establishing the mean and variance of the expected behaviour.
- *To understand project limits*: How likely is CO_2 migration towards critical project boundaries (e.g. spill points, licence boundaries) or when could pressure limits be reached (e.g. permissible limits related to fracture pressure). These questions are better address by scenario modelling of specific cases, within the context of reasonable risk (i.e. scenarios that have a physical basis and exceed a threshold).
- *To find ways to optimise the injection plan and strategy*: How could well placement and flow rates be optimised to ensure best use of the storage target in terms of best use of the available pore space and various geological units.

For all these objectives, the ability to update and adjust the injection plan as new monitoring data is acquired during injection should be an integral part of the forecasting approach.

In Sect. 2.5.4 above, we reviewed some of the many studies on modelling the CO_2 plume at Sleipner. Two general observations emerge from these studies:

- Despite the complexity of the multi-layer plume at Sleipner the overall plume footprint is consistent with the initial expectation—the plume has stayed within the intended structural closure targeted by the project (Zweigel et al. 2004).
- The excellent seismic imaging datasets have stimulated significant learnings in terms of flow simulation. In their assessment of multiple flow model approaches to matching the plume at Sleipner, Williams et al (2018) concluded that although there were minor differences related to solver implementations, grid meshing and equation-of-state uncertainties, the differences were less than the monitoring uncertainties.

Fig. 3.6 Example CO_2 plume simulations from the Snøhvit project (Tubåen reservoir). Top: CO_2 plume forecast after 20 years of injection using a conventional vertical well; Bottom: CO_2 plume forecast after 20 years of injection for a scenario using an extended reach multi-branch well. Vertical scale is exaggerated by 7; reservoir unit modelled is 80 m thick. thick (from Nazarian et al. 2013; © Elsevier, reproduced with permission)

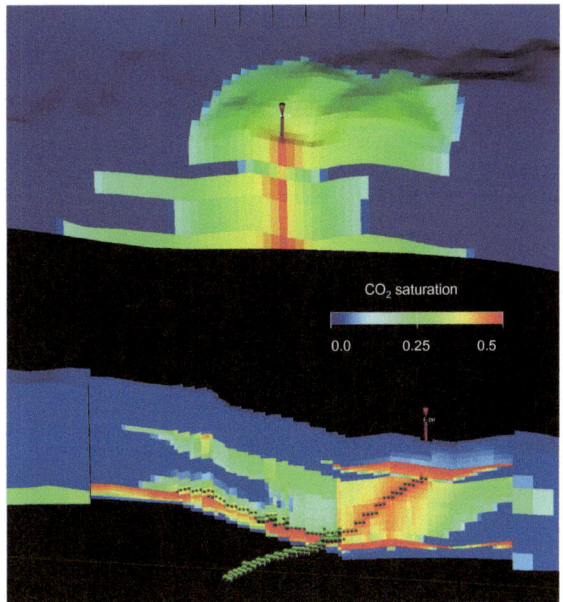

This experience should provide greater confidence for future storage projects—namely that it is possible to forecast the likely plume behaviour within reasonable bounds of uncertainty. Precise predictions cannot however be expected, and an interactive modelling approach with model updates based on feedback from regular monitoring data is the right approach for developing CO_2 storage projects.

To illustrate the benefits of forecasting using reservoir flow simulation, Fig. 3.6 compares two scenarios modelled the Snøhvit (Tubåen Formation) CO_2 injection case. The objective here was to test the benefits of an extended reach multi-branch injection well compared to conventional vertical well—the actual well used at Snøhvit. This study (Nazarian et al. 2013) showed how a horizontal well utilizes much more of the storage place available in this layered reservoir. In the conventional vertical well solution, only the layers with highest permeability contribute to CO_2 storage due to differential injectivity, while the advanced injection well allows much more effective use of the stacked reservoir units. While, advanced injection wells for CO_2 storage are currently too expensive, the study illustrates the potential for future optimization of CO_2 storage using smart well placements.

3.4.3 Using Geo-Pressure to Manage the Storage Strategy

A key concern for CO_2 storage is understanding and handling the pressure limits—CO_2 injection pressure cannot be allowed to exceed a certain rock mechanical limit (see Sect. 2.6 above). Studies of individual projects (Rutqvist 2012) and especially large-scale storage potential at the basin scale (Gasda et al. 2017; Ganjdanesh and Hosseini 2018) need to assess the expected pressure development in the project and ensure a critical maximum pressure limit, usually the caprock fracture pressure, is not exceeded. A good way to approach this problem is to consider this in terms of the basin geopressure. In Sect. 2.6 we introduced the concepts of rock mechanics and effective stress, and it is useful now to develop this into a framework for storage pressure management.

Figure 3.7 shows a generic approach to pressure management based on pressure and stress trends in an example offshore sedimentary basin (the Norwegian North Sea). Assuming the main interval of interest for storage is between 800 m and 4000 m depth, we can expect most saline aquifers to have an initial pressure close to hydrostatic pressure. In some cases, deeper units (below around 3000 m) may be over-pressured and have a higher initial pressure. However. the concept is similar—most projects will have an available pressure increase of between 5 and 20 MPa before the fracture pressure is reached. This is in contrast with oilfield developments where the initial excess pressure due to the buoyant hydrocarbon can be depleted towards the hydrostatic pressure, or lower. Depleted gas fields may have pressures substantially below hydrostatic, giving a higher available range for the allowable pressure increase (Bouquet et al. 2009; Nazarian et al. 2018).

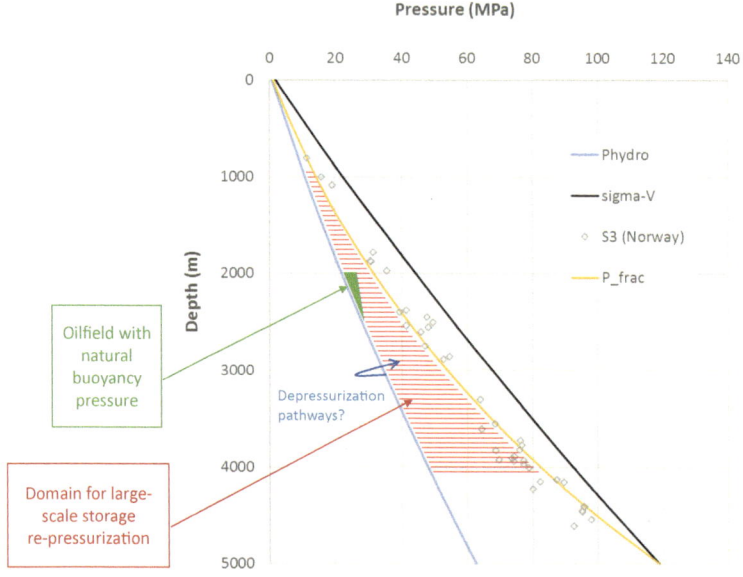

Fig. 3.7 Pressure depth functions for a generalised Norwegian North Sea basin case illustrating the domain for large-scale re-pressurisation of aquifer units during storage. Phydro is the hydrostatic gradient, sigma-V is the vertical principal stress, and the maximum reservoir pressure is described by the formation fracture pressure P_frac based on minimum stress data (S3) from Bolås and Hermanrud (2003)

Ringrose and Meckel (2019) have developed this concept further in an assessment of global CO_2 storage resources on offshore continental margins, where they propose two broad classes of storage project with contrasting operational criteria for project termination:

A. Storage projects that fill the available pore-space before the maximum pressure limit is reached (Aquifer geometry A in Fig. 3.8);
B Storage projects that reach the maximum pressure limit before the available pore-space can be fully utilized (Aquifer geometry B in Fig. 3.8).

The concept outlined in Fig. 3.8 is scaled to a common set of initial conditions: the initial reservoir pressure, P_{init}, the bottom-hole well pressure, P_{well}, and the formation fracture pressure P_{frac}. Storage geometry A follows a pressure path P_a towards a final pressure P_{fa}, and likewise for B. The Sleipner projects is an example of A, while the early injection history at the Snøhvit project is an example of B.

Of course, each project will have unique criteria, and needs to be assessed on a case by case basis; however, this basin geopressure framework helps to explain the underlying concept. As storage projects develop to the situation of multiple projects within the same basin, this geopressured setting will also be vital for assessing how pressure might evolve within the basin as a whole.

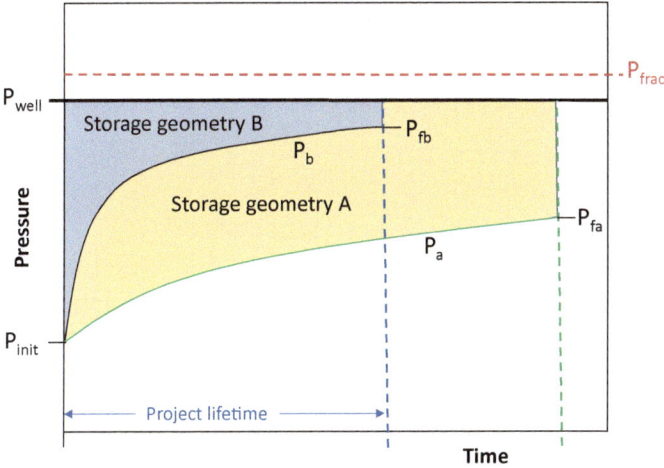

Fig. 3.8 Idealised project lifetime pressure plots for two contrasting aquifer units assuming the same initial pressure conditions

3.5   Handling Well Integrity Issues

Many analyses of leakage risks associated with storage have concluded that wells represent one of the most important leakage risk factors (Pawar et al. 2015). Man-made penetrations of geological sealing units represent a key risk factor. Studies of well integrity make a distinction between wells constructed for the specific purpose of injecting and monitoring CO$_2$ (designed appropriately) and legacy wells that exist within the site area that are either operational or abandoned. However, approaches for handling these risks have been developed (Oldenburg et al. 2009), and so the focus in CO$_2$ projects is on quantifying and handling these risks (see Zhang and Bachu 2011 for a review).

To appreciate the issues involved, note that wells are multi-component engineered systems (Fig. 3.9) and are designed with a multiple barrier system to ensure a pressure seal is maintained during and after operation. The multiple barriers comprise the tubular elements, elastomer- and metal-seals, cemented sections and the wellhead system itself (also containing multiple valves and seals). Cementation of the well to the host rock is generally focused on the important barriers within the caprock and

Fig. 3.9 Example CO$_2$ injection well design (modified from Cooper et al. 2009)

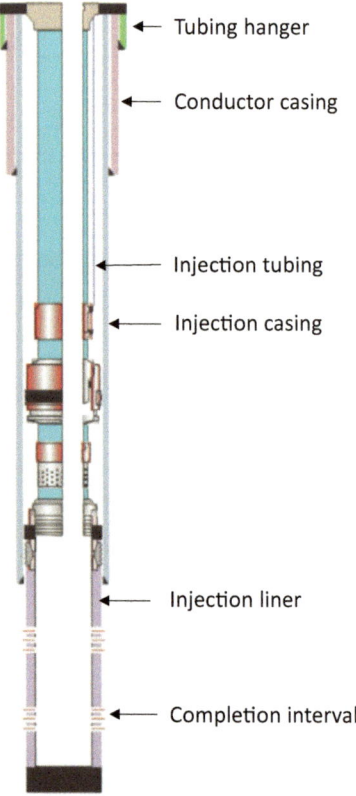

← Tubing hanger

← Conductor casing

← Injection tubing

← Injection casing

← Injection liner

← Completion intervals

shallower sealing units. Wells are also the focus of monitoring activities, both during the injection period (e.g. careful monitoring of pressures) and during the post-closure period (when a monitoring plan is generally required to ensure the barrier system is functioning).

For CO_2 storage projects two issues are foremost in the well design:

- *Handling and minimizing corrosion of metallic elements*: Corrosion is firstly limited by ensuring dehydration of the CO_2 injection stream (usually to <100 ppm H_2O), but where significant amounts of O_2, H_2S or H_2O are present in the injection stream, the well design is then adjusted by choosing corrosion resistant alloys for the relevant well components (see Sect. 3.1 above).
- *Design of cement isolation elements of the well design*: A secure well requires careful placement of cement slurry around casing at key intervals (around the casing shoes and at the surface). Most wells use Portland-based cement with various additives to alter the cure time, strength, and sulfate resistance (class G cements are most commonly used). To give better CO_2-resistant behaviour special cements may be used (such as EverCRETE™ or ThermaLock™).

In summarizing their findings on wellbore cements for CO_2 injections wells, the industry-led Carbon Capture Project (CCP; Cooper 2009) concluded that:

- Standard Portland-based cements and carbon steel casing can provide long-term hydraulic isolation;
- Emphasis should be placed on good cement placement practice (rather than special CO_2-resistant cement) to provide an effective barrier;
- Cement interfaces (not the matrix) are the most likely path for fluid migration.

In response, ongoing research has been especially focused on what could happen at the critical interfaces—cement to rock and cement to steel—looking at both thermal stresses and chemical reactions (Carroll et al. 2011, 2016; Van der Tuuk Opedal et al. 2014).

Assuming that a good well design can ensure successful CO_2 storage during operations, much of the discussion around wells has focused on what might happen after site closure—how long can the well last as an effective seal and could it leak at some point in the future? These are more difficult questions to address, and this is an arena where learning from 'early-mover projects' is extremely valuable.

In their analysis of long-term wellbore integrity at the In Salah CCS project, Carroll et al. (2011) and McNab and Carroll (2011) performed detailed geochemical modelling and experiments to study brine-CO_2 reactions with wellbore cement and caprock. They drew some general conclusions, that should help inform studies of wellbore integrity for future projects:

- Analysis of the solution chemistry and the solid products over time shows that the wellbore environment is dominated by reactions between cement, carbonate, and clay minerals when exposed to a CO_2-rich fluid.
- Reaction of the hydrated cement with synthetic brine (equilibrated with supercritical CO_2) rapidly occurs (usually within 5–10 days) to form amorphous silica,

calcite, and aragonite. Similar reaction products were observed in experiments with sandstone and CO$_2$ (representing the reservoir).

- Geochemical modelling to assess the potential impacts of these observed reactions suggests:

  – Significant retardation of the rate of advance of CO$_2$ occurs along potential interfaces (e.g. cement-rock interface)
  – Porosity reduction of a few percent is expected and there are relatively minor impacts on permeability.

To summarize these studies further, although important geochemical reactions do occur (especially between CO$_2$-rich fluids and cement), the resulting cementation processes tend to seal up fractures and pores. So, it is likely that CO$_2$ storage is a self-sealing system—because introducing CO$_2$ into a saline aquifer leads to cementation. However, the processes are complex, and it is clear that there are potential weak points in the wellbore system that need to be better understood, especially at cement interfaces.

To look ahead and prepare for what might be required when closing and completing a CO$_2$ storage project, the Ketzin CO$_2$ storage pilot site provides an excellent prototype (Fig. 3.10). The Ketzin project (Martens et al. 2012, 2014) was the first European pilot site for onshore storage of CO$_2$ in saline aquifers. A total of 67 kt of

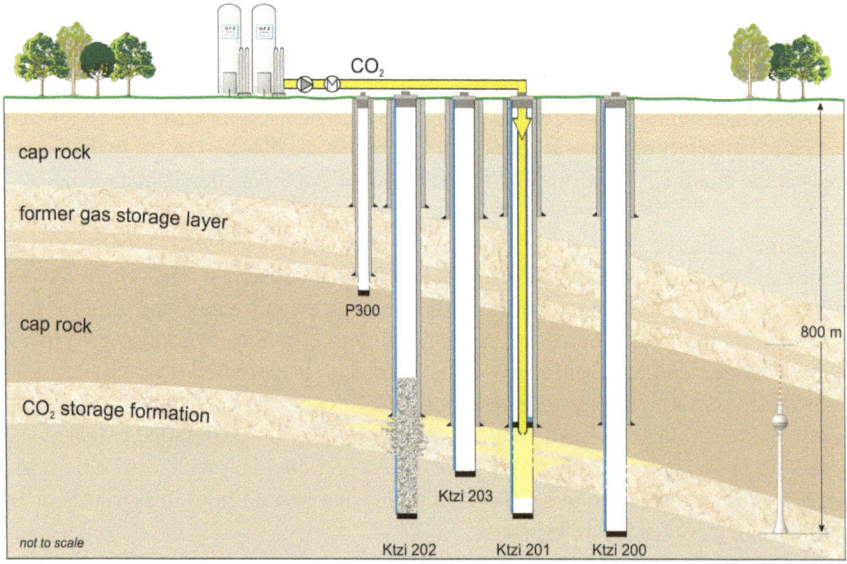

Fig. 3.10 Schematic of the Ketzin pilot site at the time of site closure (2015). Well Ktzi 201 is the injection well and the other four wells are observations wells. Partial plug and abandonment (P&A) of the observation well Ktzi202 was completed in 2013 (image from Gawel et al. 2017; © Elsevier, reproduced with permission)

CO_2 was injected between June 2008 and August 2013, when injection was stopped, and the site entered the post-closure phase.

During operation the focus was mainly on developing and testing monitoring technologies (Ivanova et al. 2012; Liebscher et al. 2013), but the fact that the project has now passed the post-closure milestone is perhaps the most valuable aspect of the project. Research and demonstration activities at site continue to address the entire life cycle of this storage site, and the post-closure phase (2013-) involves a multidisciplinary monitoring program and a stepwise abandonment of the five wells drilled at the site (Martens et al. 2014).

Being a research pilot and demonstration project at an onshore location, this project is able to demonstrate safe and reliable on-shore CO_2 storage including fulfilment of legal and regulatory requirements. The project continues to be used as a case example for developing national and international standards (DIN/ISO), and is also used in public engagement activities to help make CO_2 storage more understandable to the public.

Whatever way the well closure process is handled, the important underlying principle is that the well must have functional barriers that provide isolation:

- between important geological intervals;
- between the well annuli;
- between the wellhead and the external environment.

Good isolation requires careful placement of cement slurry around casing and at plugged intervals. Normally, this is only done at key intervals around casing shoes, sealing units and at the surface (Fig. 3.11). In some cases, cementation of the entire injection casing back to the surface may be required.

The final and most challenging issue related to well integrity is how to handle inherited wells, or legacy wells (Zhang and Bachu 2011; Pawar et al. 2015). These are typically oil and gas exploration wells which have not been completed in a way to ensure sealing of the well at higher levels. For example, after testing of a deeper oil prospect (say at Jurassic level) and finding no oil, the well was plugged at the deeper level and at the surface, while the shallower saline aquifer level was not sealed and offers a potential leakage path for future CO_2 storage projects. The impact of these legacy wells is difficult to quantify, but their existence certainly impacts the cost and availability of potential storage site developments. For CO_2 storage project developments, there are essentially two approaches to this issue:

1. Avoid them—and develop storage projects only where secure well completions are in place;
2. Quantify the leakage risk involved in each legacy well in the area of interest, and then decide if the leakage risk is low enough to proceed or if remediation of poorly sealed legacy wells can be done cost effectively.

Early mover projects, such as Sleipner, have chosen the first option; but with time we can expect future projects to include well-managed or remediated legacy wells in their development plans. It is also likely that the issues involved with legacy wells will be first addressed where depleted oil and gas fields are used for storage—since such

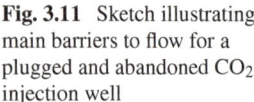

Fig. 3.11 Sketch illustrating main barriers to flow for a plugged and abandoned CO_2 injection well

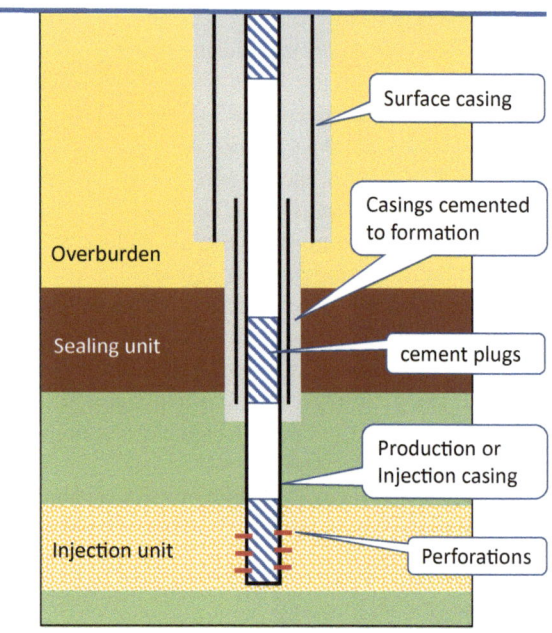

projects will inevitably have to manage legacy well issues. For example, the Lacq-Rousse CO_2 Capture and Storage demonstration pilot project in a depleted gas field in France (in operation between January 2010 and March 2013), includes an injector well which was plugged and abandoned in May 2015. A risk assessment study of potential leakage through the wellbore cement (Thibeau et al. 2017) concluded that no CO_2 would flow towards the aquifer because, in the event of leakage, the pressure gradient would lead to the flow of formation water into the CO_2 storage unit. Even in the case of leakage into the depleted pressure CO_2 storage unit, the stabilized flow rate was found to be insignificant (at around 0.01 m³/day).

3.6 Monitoring CO₂ Storage Projects and Managing Site Integrity

3.6.1 What Type of Monitoring Is Needed?

Finding the best way to monitor CO_2 storage projects has drawn a lot of attention and research focus over the last two decades, and there are now several best practice documents and textbooks on this topic to guide future projects (e.g. Chadwick et al. 2008; Davis et al. 2019). What we will focus on here is how to approach and design the monitoring plan and how to best use the latest advances in monitoring technology.

Many of the developments in petroleum reservoir monitoring can be adapted and applied to CO$_2$ storage monitoring, especially the now widespread use of time-lapse seismic reservoir monitoring in the oil industry (Lumley 2011) which is very suited to monitoring the CO$_2$ plume in the subsurface, as demonstrated at the Sleipner project (Arts et al. 2004). However, CO$_2$ storage monitoring presents us with some different challenges related to the behaviour of CO$_2$ in the subsurface and the additional concerns around ensuring safe long-term storage. In short, the physics of monitoring CO$_2$ storage is different from oilfields and the societal demands are more extensive.

A CO$_2$ storage project monitoring portfolio has to cover several issues:

- Ensure safe site operations;
- Satisfy regulatory requirements;
- Address public concerns about possible leakage;
- Ensure secure long-term storage of CO$_2$.

To meet these demands, early-mover projects have tried to develop fit-for-purpose approaches to monitoring, and the lessons learned from that process should be very valuable to future projects. In addressing the question 'What type of monitoring is really necessary?' projects need to consider three overall questions related to different stakeholder viewpoints:

1. What monitoring is important from an operational point of view?
2. What monitoring is required from a regulatory perspective?
3. What monitoring is in the public interest?

These viewpoints can lead to conflicting demands and priorities, although ultimately the monitoring plan must address all three. Another major technical challenge is that projects may potentially see the need to monitor a considerable volume of rock, that is:

Not only the reservoir (saline aquifer) target zone

… but also the overburden sequence
… and the surface environment in the region around
… and the surface facilities (pipes, wellheads, etc.)
… and a period of many years after the site closure.

These rather extensive demands create significant challenges for project prioritization decisions; and learning what monitoring activities are actually needed for each specific project will be key for widespread deployment of CO$_2$ storage.

3.6.2 Monitoring Objectives and Definitions

According to the European Storage Directive (EC 2009), the overall objective of a CO$_2$ storage site monitoring programme is to verify storage and minimize the risk of leakage. Different national jurisdictions may have slightly different wordings and legislations in place, but the objectives are similar. Note the following general terminologies used:

- Monitoring = regular observation and recording of a project or programme (the preferred term in EC CO$_2$ storage Directive)
- MMV = Measurement, Monitoring and Verification (a technical description of monitoring activities for a CO$_2$ Storage project)
- MVA = Monitoring, Verification, and Accounting (similar to MMV, but including the legal accounting aspects; the preferred term in the US NETL Best Practice).

Furthermore, MMV programmes for CO$_2$ Storage projects need to address the main project phases (refer back to Fig. 2.4):

- Pre-injection (site selection)
- Operational
- Site closure
- Post-closure.

In addressing the technical objectives of monitoring, there are two main goals which are generally defined as follows:

- *Conformance*: The process of verifying that storage performance in the subsurface proceeds as expected.
- *Containment*: Activities to ensure and verify that the injected CO$_2$ is contained within the storage complex.

The third important technical objective is *Contingency*—the ability to respond to any anomalies that are detected and if necessary to stop any leakage that could occur.

There are also various important regulatory requirements that must be met, including:

- Reporting to the competent authority at least once a year (EC Directive);
- Meeting Environmental Protection requirements, especially to protect underground sources of drinking water (US EPA regulation of USDWs) or to protect the marine environment (London and OSPAR conventions);
- Agreements for handling legal liabilities during the post-closure monitoring phase and eventual transfer of responsibility of the site back to the relevant national authority.

For further reading on legal and regulatory aspects of CO$_2$ storage see Dixon and Romanak (2015) and Dixon et al. (2015).

3.6.3 Designing the Monitoring Programme

Before discussing approaches for deciding what to include in a monitoring plan, it is useful to review choices that have been made from example projects. Table 3.1

Table 3.1 Monitoring methods applied at three industrial-scale CO$_2$ storage projects

Monitoring Technology	Sleipner (offshore platform)	In Salah (onshore)	Snøhvit (offshore subsea)
Wellhead monitoring	✓	✓	✓
Downhole fluid sampling	✓	✓	✓
4D seismic	✓	✓	✓
4D gravity	✓		✓
Seabed/marine surveys	✓		✓
Microseismic monitoring		✓	
Permanent down-hole gauges			✓
Well-testing during operations		✓	✓
Wellbore integrity monitoring		✓	✓
CO$_2$ tracers		✓	
Satellite (InSAR) monitoring		✓	
Surface/shallow gas	✓	✓	
Groundwater sampling		✓	

summarizes the main choices made for the Sleipner, In Salah and Snøhvit projects—a useful comparison since one is offshore platform based, one is onshore and the third involves a subsea injection well. In a fuller review of monitoring technologies, Jenkins et al. (2015) published a more complete list comparing these three projects with two CO$_2$-EOR and storage projects (Weyburn and Cranfield) and one research injection site (Otway).

There are several factors involved in the choices made:

- Wellhead monitoring, down-hole fluid sampling, and time-lapse (4D) seismic were used at all sites, attesting to their essential value for CO$_2$ storage projects;
- Several technologies are either only, or mainly, suitable for onshore sites, namely InSAR (Satellite airbourne radar inteferemetry) and ground-water sampling;
- Time-lapse (4D) gravity field monitoring has proven especially accurate in the offshore setting;
- Permanent downhole-gauges technology was immature for the earlier projects (Sleipner in 1996 and In Salah in 2004) but more reliable by the time Snøhvit started in 2008.

Fuller accounts of the monitoring choices made at these projects are given by Furre et al. (2017), Mathieson et al. (2010) and Hansen et al. (2013).

An important point here is that monitoring technology is continually improving and the cost gradually falling, so that future projects will be better placed to adopt mature and proven technologies. In particular, down-hole fibre-optic sensing is a rapidly developing field and likely to be widely deployed in future sites. Key technologies are Distributed Temperature Sensing (DTS) and Distributed Acoustic Sensing (DAS) for recording seismic events. The recent onshore CO$_2$ storage projects in Canada at Quest (Bourne et al. 2014; Mateeva et al. 2014) and at Aquistore (Worth et al. 2014; White et al. 2017) have demonstrated the value of time-lapse vertical seismic profiling (VSP) as a means of cost-effectively monitoring plume growth.

Extrapolating forward, based on project experience so far, we can draw a picture of what future CO$_2$ monitoring portfolios might look like (Fig. 3.12).

The following are expected to be either essential or dominant for future projects:

- Geological site characterisation datasets are essential and will typically include several wells (with an extensive logging and coring programme), surface surveys and 3D seismic surveys covering the site volume.
- Standard wellhead and downhole measurements (regular or continuous measurement of pressure, temperature and fluid composition).

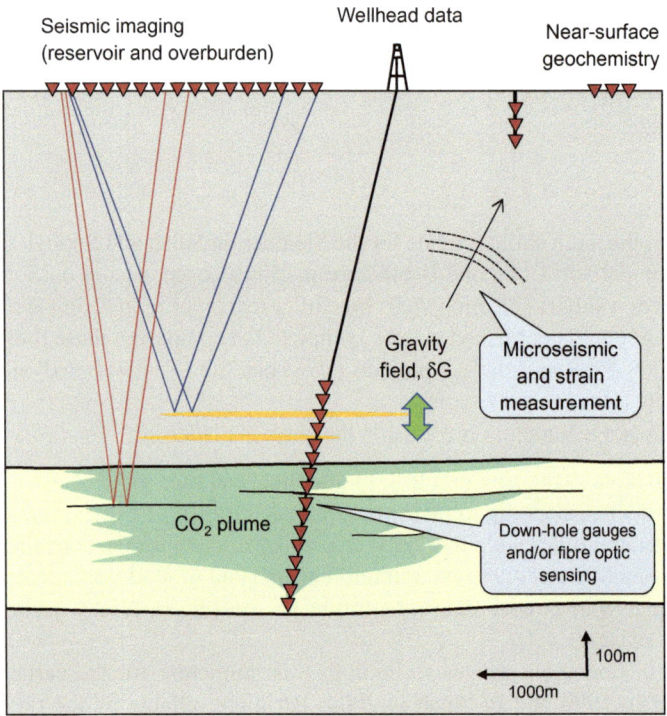

Fig. 3.12 Idealised sketch of a storage site monitoring programme

		OSPAR	EU Directive	EU ETS	
Deep-focussed monitoring actions	Migration in overburden				Containment
	Containment integrity				Containment
	Migration in reservoir				Conformance
	Performance testing and calibration and identification of irregularities				Conformance
	Calibration for long-term prediction				Conformance
	Testing remedial actions				Contingency
Shallow-focussed monitoring actions	Verification of no leakage				Containment
	Leakage detection				Containment
	Emissions quantification				Contingency
	Environmental impacts				Other
	Testing remedial actions				Contingency

Fig. 3.13 Monitoring choices to address regulatory requirements in the offshore setting (from Hannis et al. 2017)

- Time-lapse (4D) seismic monitoring (with many options on the type of seismic acquisition and repeat intervals).
- Distributed fibre-optic sensing (DAS and DTS)—both downhole and at surface.
- Monitoring rock strain and microseismic events, using either purposed arrays of 3-component geophones or surface deformation monitoring (e.g. using satellite InSAR datasets).
- Gravity field monitoring, especially for larger offshore projects.
- Surface gas monitoring (with quite different strategies for onshore and offshore settings).

Another important perspective on monitoring choices is the question of 'what might be required from a regulatory perspective?' This question has been recently reviewed for the offshore setting by IEAGHG (2016) and Hannis et al. (2017). Their analysis differentiates shallow- and deep- focused monitoring and how the monitoring action address requirements related to conformance, containment and contingency (Fig. 3.13). In general, regulatory and public interests tend to prioritize monitoring of the shallow or surface environment while operational concerns are more focused on the deeper setting at or close to the storage unit. A balance of objectives is required.

3.6.4 Monitoring Insights from Selected Projects

To illustrate actual monitoring choices made for selected projects it is useful to summarise what was done at the In Salah project (onshore) and the Sleipner project offshore. Figure 3.14 summaries the monitoring programme at the In Salah project

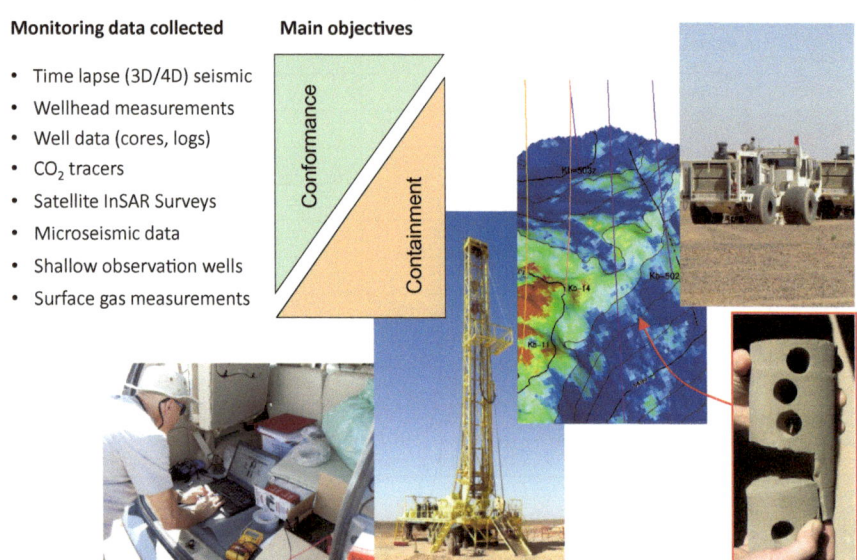

Monitoring data collected   **Main objectives**

- Time lapse (3D/4D) seismic
- Wellhead measurements
- Well data (cores, logs)
- CO$_2$ tracers
- Satellite InSAR Surveys
- Microseismic data
- Shallow observation wells
- Surface gas measurements

Fig. 3.14 Summary of storage site monitoring programme at the In Salah project. Images show (from left to right) microseismic data recording, shallow well drilling, seismic interpretation and repeat acquisition, and core analysis (inset)

(reported in detail by Mathieson et al. 2010 and Ringrose et al. 2013). Here, conformance monitoring was dominated by time-lapse seismic, well data and wellhead measurements and the use of artificial CO$_2$ tracers (to evaluate CO$_2$ breakthrough to gas production wells). Containment monitoring was focussed on measurements at shallow groundwater observation wells and surface gas measurements. Novel use of satellite radar interferometry (InSAR) monitoring datasets was used to track the growth of the subsurface pressure field (Vasco et al. 2010) and passive seismic monitoring was used to analyze injection related microseismic events (Goertz-Allmann et al. 2014), as explained in Sect. 2.6.3 above.

To assist in deciding which monitoring technologies to deploy at the In Salah site, the project used a 'Boston Square' cost-benefit assessment framework (Mathieson et al. 2010). No one wants to use high- cost/low-benefit technologies—so after numerous studies on the technical benefits of each technology the project ended up with a set of low-cost/high-benefit technologies (Fig. 3.15). However, certain high-cost technologies were deployed because of their enormous benefits. 3D and 4D seismic surveys fall into this category—high budget items which are essential. Consequently, most CO$_2$ projects spend their efforts not on whether they will acquire seismic data, but how the survey costs and number of repeats can be optimized.

Moving to our second example of the monitoring strategy adopted at Sleipner (Fig. 3.16), we see a useful assessment of the value of the number of repeat surveys taken over a 20-year period (Furre et al. 2017). The seismic monitoring programme at Sleipner has so far comprised nine repeat 3D towed-streamer seismic surveys,

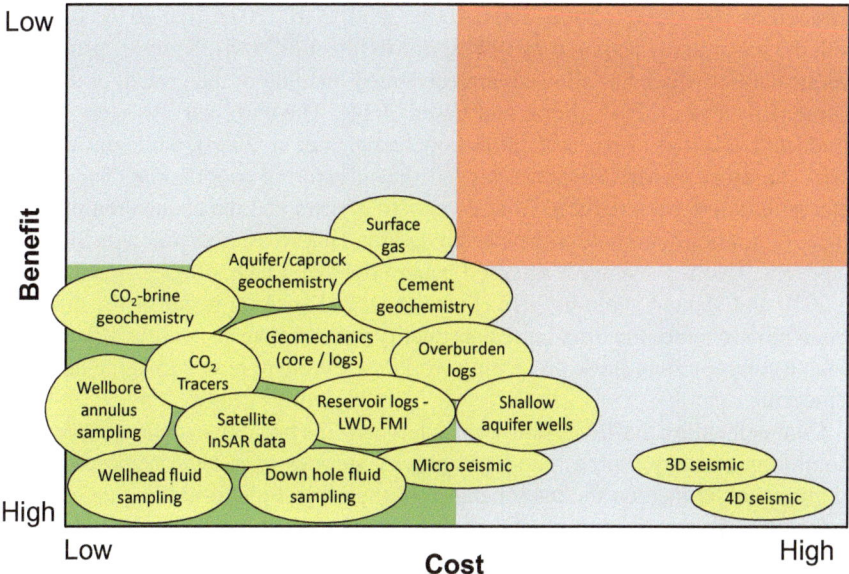

Fig. 3.15 Summary of the monitoring portfolio choices implemented at the In Salah CO$_2$ storage project, plotted on the 'Boston square' framework used for cost-benefit assessments (modified from Ringrose et al. 2013)

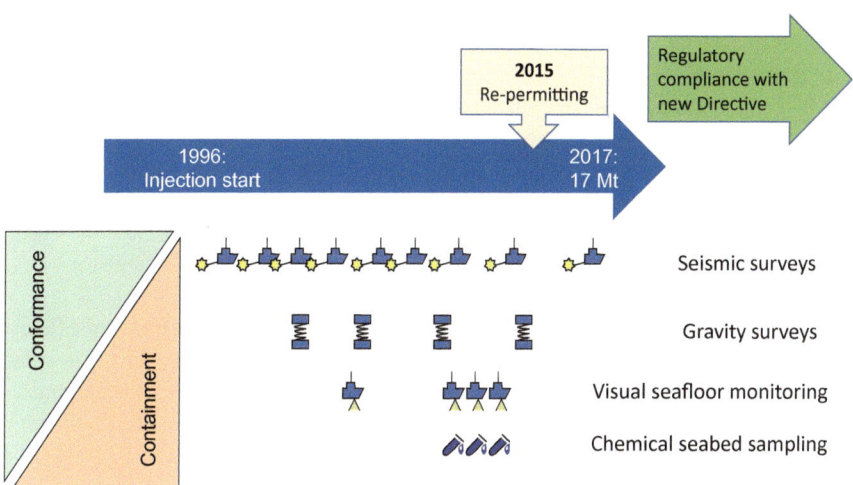

Fig. 3.16 Summary of storage site monitoring programme at the Sleipner project (modified from Furre et al. 2017)

acquired in 1999, 2001, 2002, 2004, 2006, 2008, 2010, 2013, and 2016. Together with the base survey acquired in 1994 (prior to injection start) the series provides a unique dataset which has allowed unprecedented imaging of the growth of the CO$_2$ plume (Eiken et al. 2011; Furre and Eiken 2014). The early surveys were mainly conducted as a part of research projects (Chadwick et al. 2008), while during later years the asset owner (Sleipner) has led the monitoring programme. The survey interval has also been reduced from 2 years to 3 years and the acquisition plan has varied (e.g. when combined with other surveys) and improved over time as technology improves. Two surveys were acquired using broadband technologies—dual source in 2010 and slanted cable in 2013 (Furre et al. 2017). As a result of the different acquisition schemes the time-lapse data quality varies between datasets, although by using a common time-lapse processing workflow excellent plume imaging has been achieved.

Complementing the 4D seismic (which has been the backbone of the conformance monitoring strategy), three time-lapse gravity surveys have been acquired giving additional assurance on the mass balance of the project and providing an estimate of the mean in situ density of CO$_2$ in the storage unit and an upper bound for the CO$_2$ dissolution rate (Alnes et al. 2011). Containment monitoring has been limited to seafloor visual inspections and sediment samples to confirm absence of anomalies, although the 4D seismic dataset also contributes to containment monitoring.

Another interesting aspect of the Sleipner project was that the CO$_2$ injection project was initially permitted under Norwegian Petroleum Law. Subsequently, in 2015, after the Norwegian legal framework was updated to be consistent with the EU directive on CO$_2$ storage (EC 2009), the CO$_2$ storage operation at the site was re-permitted to be consistent with the new directive (Fig. 3.16). In fact, the pioneering Sleipner project was initially used to inform the terms of the EU directive on CO$_2$ storage and was then re-permitted under the new law.

The repeat seismic surveys have been the main method for demonstrating both containment and conformance at Sleipner. The high porosity sandstone at 800–1000 m depth results in a strong seismic amplitude change when CO$_2$ enters each layer (due to the significant change in seismic velocity and effective sandstone density compared to the brine-filled reference). Detailed analysis of the thin CO$_2$ layers, using both amplitude-change and time-shift data, allows imaging of multiple layers within the plume and estimation of layer thicknesses down to a few meters, in some cases (Furre et al. 2015). Figure 3.17 shows example seismic and gravity monitoring datasets from the Sleipner project. Both datasets have been vital in confirming successful storage at this site. From the operator perspective (Furre et al. 2017), the geophysical datasets confirm that the CO$_2$ is contained in the intended target layers and structural closure. From a technology development perspective (e.g. Furre et al. 2015; Landrø and Zumberge 2017; Chadwick et al. 2019), the datasets have led to significant improvements in the technology for detection of thin CO$_2$ layers in storage units and in understanding the physics of CO$_2$ storage.

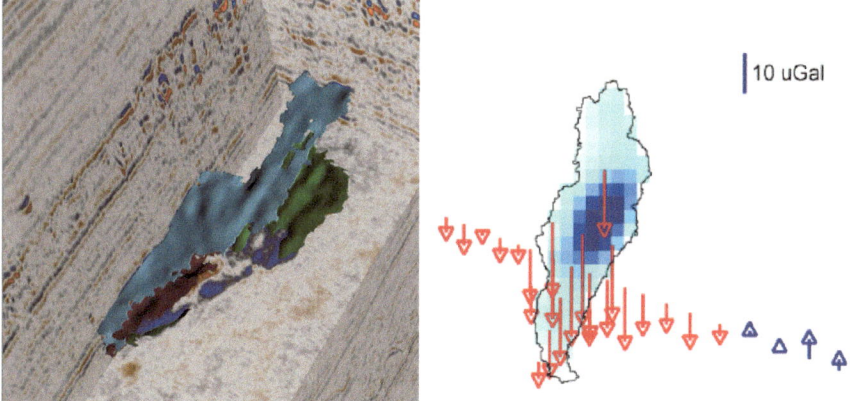

Fig. 3.17 Example monitoring datasets from the Sleipner project. Left image: Multiple CO$_2$ layers inferred from amplitude changes seen in the 2013 seismic survey (viewed from above, with layer 9 in pale blue). Right image: Total CO$_2$ thickness change based on inversion of the change in gravity field from 2002 to 2013. Red arrows show decrease in the measured gravity field and blue arrows an increase. Maximum CO$_2$ thickness change is c. 35 m (deep blue shading). Left image courtesy of Anne-Kari Furre, Equinor, published with permission of the Sleipner Production Licence. Right image modified from Furre et al. 2017; © Elsevier, reproduced with permission

Typical questions posed regarding the seismic monitoring of CO$_2$ storage include:

- How much CO$_2$ can you detect?
- And what about the CO$_2$ you cannot detect.

The Ketzin pilot injection project in Germany provides a valuable example of detecting relatively small volumes. At this site, about 67 kt of gas-phase CO$_2$ was injected into the sandstone units at a depth of 630–650 m through the injection well Ktzi 201 (ref. Fig. 3.10). The injection period was June 2008 to August 2013. The first repeat seismic survey taken in 2009 revealed a clear amplitude response after 23 kt had been injected (Fig. 3.18; Lüth et al. 2015). The second repeat seismic survey taken in 2012 successfully imaged the expanded plume after 61 kt of injection. Thus, detection of around 20 kt of CO$_2$ at this depth is clearly possible both initially and as the plume grows. In terms of the minimum thickness of the CO$_2$ layer at Ketzin that can be detected from 4D seismic, this is inferred to be around 7 m or greater (Huang et al. 2018).

Because the Ketzin site is one of the few projects to have moved into the post-closure monitoring phase, the monitoring team was able to assess stabilization of the CO$_2$ plume using a post-closure survey (Fig. 3.18; Lüth et al. 2015; Huang et al. 2018). This is not a straight-forward process, since at the time of closure CO$_2$ will continue to dissolve into the brine phase (a stabilization process in itself) and may continue to migrate laterally. At Ketzin, CO$_2$ was injected into a layer dipping at around 15° on the southern flank of the Ketzin anticline. The 2015 post-closure seismic survey confirmed the overall stability of the detected plume but revealed some reduction in the size of the plume, probably due to both the dissolution effect and migration

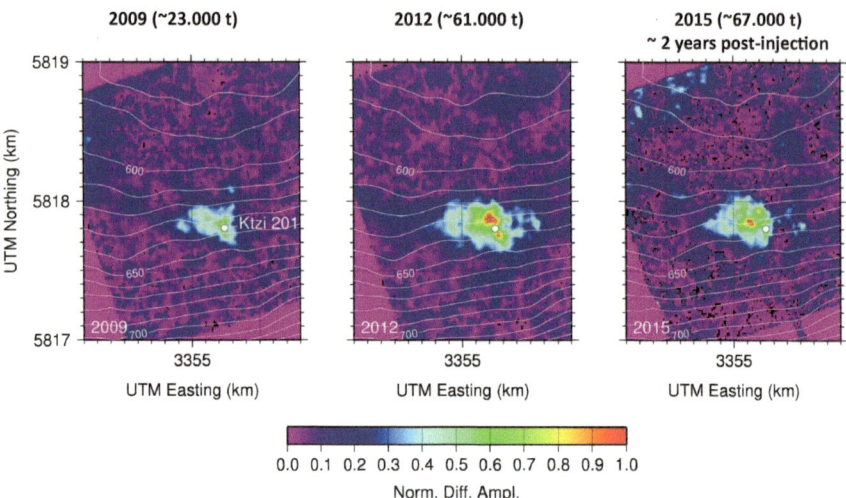

Fig. 3.18 Normalized amplitude-difference maps of the storage layer at the Ketzin site from three repeat surveys, the last being 2 years post-injection (modified from Lüth et al. 2015; © Elsevier, reproduced with permission)

into thin layers below the detection limit. These are important findings, since future projects will need to determine what level of assessment is needed to assure the relevant regulatory body that the storage site has reached a stable state. Since the legal obligation is to assure conformance and containment, technical arguments will be needed to show that the plume is approaching a stable state within the storage unit. Physical and chemical processes will continue to operate after site closure, but these should be on the path to increasing storage security over time (Ref. Figure 2.5).

We have so far mainly focussed on geophysical methods for monitoring; however, direct measurements within and around the storage complex are also important. Two groups of measurements are important here:

- Down-hole well measurements—mainly pressure, temperature and fluid saturation;
- Measurements in the surface and near-surface environments—mainly methods for gas detection and geochemical monitoring of ground water systems.

The potential for use of multiple downhole measurements, is very well illustrated by the Illinois-Basin Decatur Project (IBDP) in the USA, a CCS pilot project which ran from November 2011 to August 2014 injecting 1Mt of CO$_2$ during that period (Finley 2014; Gollakota and McDonald 2014). CO$_2$ at this site was injected near the base of the Cambrian Mt Simon Sandstone (between 2129 and 2138 m deep) into a fluvial sandstone with porosities of 18–25% and permeability in the of 40 mD to 380 mD. During this period, the project collected a unique set of data from the injection well (CCS1) and a monitoring well (VW1), including downhole temperatures and pressures at multiple depths—below the reservoir, eight zones within the reservoir, and two zones above the reservoir (Couëslan et al. 2014). The project also recorded microseismicity using surface and downhole geophones. An example of data collected over a 3-month period of sustained injection is shown in Fig. 3.19, displaying key injection parameters of downhole pressures, injection flow rate (tonnes/hour) and magnitude of recorded microseismic events (Moment Magnitudes). Real-time downhole pressures were collected at 11 zones in the VW1 monitoring well, and data from two zones (Zones 3 & 4) are displayed in Fig. 3.19. The observed differences in pressure response recorded in the multiple levels provide useful insights into the distribution of pressure within the reservoir system. Note that the Zone 3 pressure closely tracks the injection pressure (in hydraulic communication) while Zone 4 (only 33 m above) has a fully muted response. This dataset illustrates an important principle for CO₂ storage monitoring, namely the value of 'above zone monitoring.' By setting gauges and detectors in geological units above the target storage unit, projects can achieve early warning of unexpected pressure communication or fluid flow out of the target injection interval. These approaches were also successfully demonstrated at the Cranfield CO₂ injection test site in Mississippi, USA (Hovorka et al. 2011; Kim and Hosseini 2014).

The microseismic monitoring at the IBDP project, using both surface and downhole geophones, has also proven very valuable in understanding the relationship between injection rates, pressure distribution within and beyond the reservoir, and rock strain. Here, a dedicated geophysical monitoring well (GM1) with a 31-geophone array at 624–943 m depth, as well as geophones placed in the injection well (CCS1), were used for monitoring the microseismic activity (Will et al. 2016). 'Listening to the rock' close to the injection interval (namely microseismic monitoring) is also emerging as a useful approach to assuring safe operations and long-term

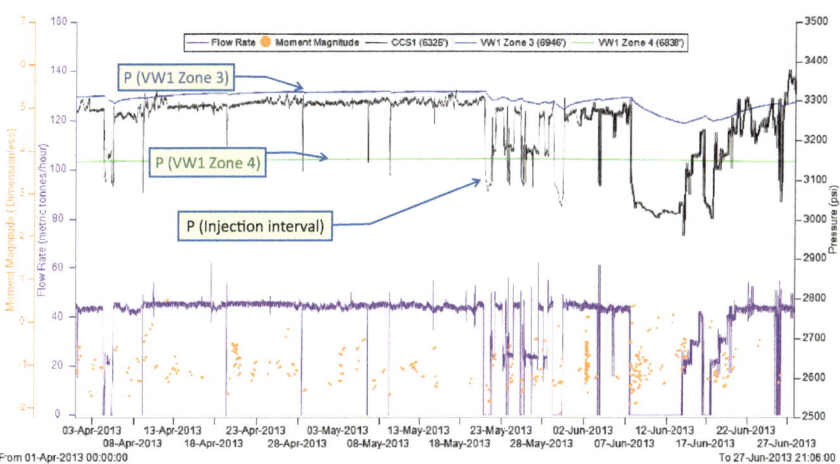

Fig. 3.19 Pressure variations at different levels in the reservoir at the IBDP CO₂ injection project, along with injection flow rate and moment-magnitude (micro-seismicity) data. From Ringrose et al. 2017; © Elsevier, reproduced with permission

permanence of storage. Events recorded at this site are mostly much less than $M = 0$ (Moment Magnitude) with a few events up to $M = 1$.

The final important class of monitoring measurements needed for CO_2 projects is a whole suite of possible measurements at or near the surface. This is often called environmental monitoring, and is generally focussed on leakage detection, especially potential contamination of shallow potable groundwater resources. Although concerns about possible CO_2 leakage are legitimate and important, there are many mistaken concepts about possible leakage detection. CO_2 is a common naturally occurring molecule, and identification of 'leaked CO_2' from a naturally varying background signal is actually very difficult:

- CO_2 is biologically produced in soils and derived from root respiration and decay of organic matter (aerobic microbial respiration).
- Deeper natural sources of CO_2 can come from degassing of groundwater (containing atmospheric CO_2) or from release of organic carbon.

These processes mean that CO_2 concentrations fluctuate following daily or seasonal cycles and long-term trends. However, assurance of secure containment is needed, and many early-mover and CO_2 injection pilot projects have been working on how to detect possible leakage or how to demonstrate absence of leakage. Jones et al (2015) provide a very useful review of the work done on understanding potential environmental impacts of CO_2 leakage from geological storage sites. Because of the complexity of CO_2 occurrence in the natural environment, most of the focus is now on process-based analysis (Romanak et al. 2012) and the use of isotopic signatures for differentiating natural CO_2 from captured CO_2 (Johnson et al. 2009; Mayer et al. 2015).

To help gain an introduction to the rather extensive topic of environmental monitoring for CO_2 storage it is helpful to summarize the main methods used:

Direct detection of CO_2:

- The Infrared Gas Analyzer (IRGA) is a commonly used device to measure CO_2 concentration in the atmospheric or soil. The measurement is based on light absorption in the near infrared part of the spectrum, typically at 4.26 μm (Oldenburg et al 2003);
- An Accumulation Chamber (AC) can also be used to measure soil CO_2 flux (also measured using an IRGA);
- Eddy covariance (EC) is a technique where atmospheric CO_2 concentrations are recorded at a specific height above the ground (measured by an IRGA). Data are integrated with meteorological data to estimate a gross conservation of energy and mass from which the net CO_2 flux can be derived;
- Light Detection and Range Finding (LIDAR) uses laser radiation to probe the atmosphere and to measure trace atmospheric gases (e.g. NO_2, O_3, H_2O, CH_4, CO_2).

Geochemical characterization of CO_2:

- Basic chemistry of gas and groundwater samples (e.g. ratios of O_2 vs. CO_2) can be used to determine the likely origin of CO_2 either in gas-phase samples or of gases dissolved in the brine phase;
- Measurements of the Carbon isotopic composition of CO_2;—note that:

  - $\delta13C$ is the deviation in parts per thousand (‰) of $^{13}C/^{12}C$ ratio against a reference
  - (^{13}C is a stable isotope mainly controlled by earth-system and biological processes)
  - $\delta14C$ is the deviation in parts per thousand (‰) of $^{14}C/^{12}C$ ratio against a reference
  - (^{14}C is the longest-lived radioisotope of CO_2 also used for dating)
  - Most isotopes are measured using a standard mass spectrometer, while ^{14}C is measured using an accelerator mass spectrometer (AMS).

- Nobel gas chemistry (e.g. $CO_2/^3He$ ratios) where the Nobel gas content is used as a natural tracer or process diagnostic tool.

The use of Carbon-isotopic and Noble-gas signatures is a particularly promising means of detecting and differentiating different sources of CO_2. The approach was successfully demonstrated in the assessment of an alleged CO_2-leakage incident linked to the Weyburn-Midale CO_2 monitoring and storage project in Canada (Gilfillan et al. 2017).

Gilfillan et al. (2014) give a summary of the progress made over the past decade in using noble gases and stable carbon isotopes as a tracing technique in CO_2 storage studies. An example diagnostic plot is show in Fig. 3.20. Note that since the only subsurface source of $^{20}Neon$ is the formation water, the Neon concentration indicates a gas phase that has been in contact with formation water. The correlation between the decreasing $CO_2/^3He$ ratio and an increasing ^{20}Ne concentration is very clear and

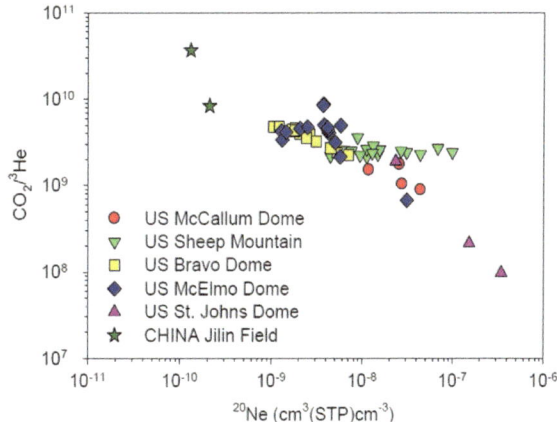

Fig. 3.20 Plot showing $CO_2/^3He$ variations plotted against ^{20}Ne for samples from data set of CO_2-rich natural gas fields (from Gilfillan et al. 2014)

can be quantitatively related to the volume of formation water that has been contacted or seen by the CO_2 gas in the geosphere.

3.6.5   Future Directions in Monitoring

In the previous section we have highlighted some of the key insights gained on monitoring CO_2 storage sites, based on early-mover saline aquifer storage projects. Some of the techniques tested may not in the long term prove to be needed, and all of the technologies can be expected to mature and develop in the coming decade. Assuming that CCS technology grows into a significant activity as part of the drive towards future low-Carbon energy systems, we can expect monitoring systems to converge towards a highly optimized and cost-effective system for assuring successful isolation of CO_2 from the atmosphere.

The techniques and approaches available for monitoring are rapidly evolving, especially in the domain of digital sensing and remotely-operated detectors. Fibre-optic based detection and signal transmission is a rapidly evolving theme, and several projects have already demonstrated the value of downhole distributed acoustic (DAS) sensing for time-lapse monitoring of a CO_2 plume (Mateeva et al. 2014; White et al. 2017).

In considering the case for optimized monitoring in the offshore setting (Fig. 3.21), Ringrose et al. (2018) assess various ways of developing optimised monitoring systems. In particular:

- How could the use of conventional marine-streamer seismic surveys be optimised?
- How could sparse permanent seabed arrays be used to best effect?
- What could be the optimal deployment of downhole monitoring devices (electric and fibre)?
- How could marine environmental monitoring surveys be best targeted and optimised?

The main solutions suggested to achieve this optimization were (Fig. 3.22):

1. To use a limited marine-streamer seismic acquisition plan (i.e. less frequent than historically used at the Sleipner project), supplemented by a sparse array of ocean bottom nodes (OBN) giving more frequent (semi-permanent) monitoring of the immediate storage site volume;
2. Use of fibre-optic downhole monitoring wherever possible, including fibre deployed pressure gauges, Distributed Temperature Sensing (DTS) and Distributed Acoustic Sensing (DAS);
3. Environmental monitoring using purposed autonomous underwater vehicles (AUV) surveys, with survey plans targeted on key points of concern (e.g. abandoned wellheads, pipelines and natural seepage sites);
4. Use advanced data analysis methods to get more value out of monitoring the data (especially full-waveform inversion of seismic monitoring datasets);
5. Development of 'trigger-survey' concepts, namely additional surveys to be scheduled only when an anomaly is detected.

Fig. 3.21 Sketch of monitoring system for CO$_2$ injection from a sub-sea wellhead into a target storage formation (green) c. 1200 m deep. Concepts illustrated include towed-marine streamer seismic acquisition, an array of seabed nodes (black dots), downhole fibre-optic monitoring (not shown) relayed via umbilical cable to shore (black line) and use of autonomous underwater vehicles (AUVs; illustrated by sub-sea illumination lighting). (Image courtesy of Equinor)

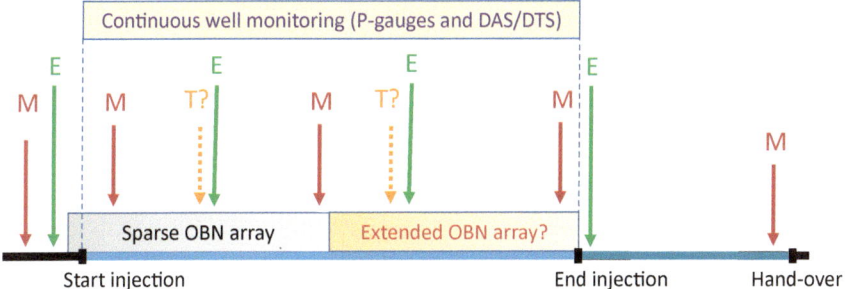

Fig. 3.22 Concepts for optimising monitoring plans for offshore storage projects. M = marine seismic survey; E = marine environmental survey; T? = optional trigger survey; OBN = ocean bottom node

Exactly how monitoring technologies will develop in the future is hard to predict; but at a fundamental level, monitoring systems should be used both to optimize a CO_2 injection operation and to assure stakeholders that the project is proceeding safely. Monitoring should not be seen as a demand (or a cost) on a project, it should be considered as a beneficial activity ensuring an overall cost-benefit for the lifetime operation of the storage project. This study of offshore monitoring (Ringrose et al. 2018) also estimated typical costs of the monitoring programme, based on the historical experience at Sleipner and Snøhvit but assuming an idealised set of parameters. The lifetime monitoring cost was estimated to be of order 2 €/t (for a 2015 reference).

3.6.6 Site Integrity and Risk Management

Even assuming you have a well-managed CO_2 injection project in operation at a well-characterised site, there will always be some residual risks you cannot anticipate. Indeed 'surprises' might be quite common, but those surprises do not need to imply significant risk. In their review of three industrial scale projects, Eiken et al. (2011) observed that the actual CO_2 plume development was strongly controlled by geological factors that were only learned about during injection. Each CO_2 injection project will encounter 'events' (such as points of CO_2 detection or pressure changes) that need to be understood and accommodated within the injection plan. These fall under the themes risk assessment and risk management.

In responding to these challenges, the IEAGHG (2009) report proposed a risk assessment and management framework for CO_2 storage projects, which has three essential stages:

1. Assess the source of the risks
2. Evaluate risk exposure at the specific site
3. Manage the risks (an activity strongly linked with the monitoring programme).

Using this framework, Pawar et al. (2015) reviewed the experience in risk management for real CO_2 storage projects, classifying these in four categories:

- Site performance risks
- Containment risks
- Public perception risks
- Market failure risks.

Interestingly, they noted that project experience so far is evidently bringing the site performance risks and containment risks down. That is, we are learning how to execute CO_2 storage projects safely and how to bring the risk factors down to an acceptable level. Unfortunately, the public-perception risks and market-failure risks are still too high and have been the dominant cause of project failure. The challenge for future projects is therefore as much about how to communicate the risks of CO_2 storage as it is about understanding those risks.

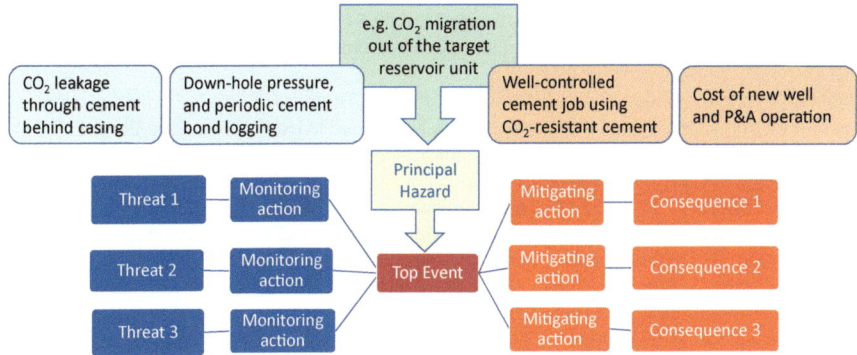

Fig. 3.23 Illustration of the bow-tie approach for practical risk management, a system to help project teams to go through potential threats and to match them to monitoring and mitigating actions. A typical example is given in the top boxes

Several projects, notably the Quest CCS project in Canada (Bourne et al. 2014; Pawar et al. 2015), have used the 'bow-tie' approach for practical risk management. A simple illustration of this approach is shown in Fig. 3.23. In practice, real projects will identify many threats with each one matched to a set of safeguards put in place to reduce the likelihood (left-hand side of the bow tie) and consequences (right-hand side of the bow tie) of any unexpected loss of containment. The Quest project (Bourne et al. 2014), also separated passive safeguards (systems that are always in place) from additional active safeguards (control measures that are triggered when anomalies are detected in the monitoring programme).

In conclusion, there is no reason to doubt that CO$_2$ storage projects can be executed safely—we have over 20 years of operational experience and have developed the tools and methods needed to control and monitor CO$_2$ storage projects and ways to handle and mitigate risks. The real challenges, at present, are the societal and market drivers needed to accelerate the technology and increase the number of projects from the current portfolio of 19 large-scale industrial scale projects (GCCSI 2019).

References

Alnes H, Eiken O, Nooner S, Sasagawa G, Stenvold T, Zumberge M (2011) Results from Sleipner gravity monitoring: Updated density and temperature distribution of the CO$_2$ plume. Energy Procedia 4:5504–5511

Arts R, Eiken O, Chadwick A, Zweigel P, Van der Meer L, Zinszner B (2004) Monitoring of CO$_2$ injected at Sleipner using time-lapse seismic data. Energy 29(9):1383–1392

Bachu S, Watson TL (2009) Review of failures for wells used for CO$_2$ and acid gas injection in Alberta, Canada. Energy Procedia 1(1):3531–3537

Bentley M, Ringrose P (2017) Future directions in reservoir modelling: new tools and 'fit-for-purpose' workflows. In: Geological society, London, petroleum geology conference series, vol 8. Geological Society of London, pp PGC8–40

Bickle M, Chadwick A, Huppert HE, Hallworth M, Lyle S (2007) Modelling carbon dioxide accumulation at Sleipner: implications for underground carbon storage. Earth Planet Sci Lett 255(1–2):164–176

Boait FC, White NJ, Bickle MJ, Chadwick RA, Neufeld JA, Huppert HE (2012) Spatial and temporal evolution of injected CO_2 at the Sleipner Field, North Sea. J Geophys Res Solid Earth 117(B3)

Bolås HMN, Hermanrud C (2003) Hydrocarbon leakage processes and trap retention capacities offshore Norway. Pet Geosci 9(4):321–332

Bouquet S, Gendrin A, Labregere D, Le Nir I, Dance T, Xu QJ, Cinar Y (2009) CO2CRC Otway Project, Australia: parameters influencing dynamic modeling of CO_2 injection into a depleted gas reservoir. In: Offshore Europe. Society of Petroleum Engineers

Bourne S, Crouch S, Smith M (2014) A risk-based framework for measurement, monitoring and verification of the Quest CCS Project, Alberta, Canada. Int J Greenhouse Gas Control 26:109–126

Carroll SA, McNab WW, Torres SC (2011) Experimental study of cement-sandstone/shale-brine-CO_2 interactions. Geochem Trans 12(1):9

Carroll S, Carey JW, Dzombak D, Huerta NJ, Li L, Richard T, Um W, Walsh SD, Zhang L (2016) Role of chemistry, mechanics, and transport on well integrity in CO_2 storage environments. Int J Greenhouse Gas Control 49:149–160

Chadwick A, Arts R, Bernstone C, May F, Thibeau S, Zweigel P (2008) Best practice for the storage of CO_2 in saline aquifers-observations and guidelines from the SACS and CO2STORE projects, vol 14. British Geological Survey

Chadwick A, Williams G, Falcon-Suarez I (2019) Forensic mapping of seismic velocity heterogeneity in a CO_2 layer at the Sleipner CO_2 storage operation, North Sea, using time-lapse seismics. Int J Greenhouse Gas Control 90:102793

Chapoy A, Nazeri M, Kapateh M, Burgass R, Coquelet C, Tohidi B (2013) Effect of impurities on thermophysical properties and phase behaviour of a CO_2-rich system in CCS. Int J Greenhouse Gas Control 19:92–100

Cooper C (ed) (2009) A technical basis for carbon dioxide storage: London and New York. Chris Fowler Int 3–20. http://www.CO2captureproject.org/

Couëslan ML, Butsch R, Will R, Locke RA II (2014) Integrated reservoir monitoring at the Illinois Basin–Decatur Project. Energy Procedia 63:2836–2847

Dake LP (2001) The practice of reservoir engineering, revised edn, vol 36. Elsevier

Davis TL, Landrø M, Wilson M (eds) (2019) Geophysics and geosequestration. Cambridge University Press

De Visser E, Hendriks C, Barrio M, Mølnvik MJ, de Koeijer G, Liljemark S, Le Gallo Y (2008) Dynamis CO_2 quality recommendations. Int J Greenhouse Gas Control 2(4):478–484

Dixon T, Romanak KD (2015) Improving monitoring protocols for CO_2 geological storage with technical advances in CO_2 attribution monitoring. Int J Greenhouse Gas Control 41:29–40

Dixon T, McCoy ST, Havercroft I (2015) Legal and regulatory developments on CCS. Int J Greenhouse Gas Control 40:431–448

EC (2009) Directive 2009/31/EC of the European Parliament and of the Council of 23 April 2009 on the geological storage of carbon dioxide and amending Council Directive 85/337/EEC, European Parliament and Council Directives 2000/60/EC, 2001/80/EC, 2004/35/EC, 2006/12/EC, 2008/1/EC and Regulation (EC) No 1013/2006

Eiken O, Ringrose P, Hermanrud C, Nazarian B, Torp TA, Høier L (2011) Lessons learned from 14 years of CCS operations: Sleipner, In Salah and Snøhvit. Energy Procedia 4:5541–5548

Eldevik F, Graver B, Torbergsen LE, Saugerud OT (2009) Development of a guideline for safe, reliable and cost-efficient transmission of CO_2 in pipelines. Energy Procedia 1(1):1579–1585

Fanchi JR (2005) Principles of applied reservoir simulation. Elsevier

Finley RJ (2014) An overview of the Illinois Basin–Decatur project. Greenhouse Gases Sci Technol 4(5):571–579

Furre A-K, Eiken O (2014) Dual sensor streamer technology used in Sleipner CO_2 injection monitoring. Geophys Prospect 62(5):1075–1088

Furre AK, Kiær A, Eiken O (2015) CO_2-induced seismic time shifts at Sleipner. Interpretation 3(3):SS23–SS35. https://doi.org/10.1190/INT-2014-0225.1

Furre AK, Eiken O, Alnes H, Vevatne JN, Kiær AF (2017) 20 years of monitoring CO_2-injection at Sleipner. Energy Procedia 114:3916–3926

Furre A, Ringrose P, Santi AC (2019) Observing the invisible—CO_2 Feeder Chimneys on seismic time-lapse data. In: 81st EAGE Conference and Exhibition 2019

Ganjdanesh R, Hosseini SA (2018) Development of an analytical simulation tool for storage capacity estimation of saline aquifers. Int J Greenhouse Gas Control 74:142–154

Gasda S, Wangen M, Bjørnara T, Elenius M (2017) Investigation of caprock integrity due to pressure build-up during high-volume injection into the Utsira formation. Energy Procedia 114:3157–3166

Gawel K, Todorovic J, Liebscher A et al (2017) Study of materials retrieved from a Ketzin CO_2 monitoring well. Energy Procedia 114:5799–5815. https://doi.org/10.1016/j.egypro.2017.03.1718

GCCSI (2019) GCCSI CO2RE database: 2019. Global CCS Institute. https://co2re.co

Gilfillan S, Haszeldine S, Stuart F, Gyore D, Kilgallon R, Wilkinson M (2014) The application of noble gases and carbon stable isotopes in tracing the fate, migration and storage of CO_2. Energy Procedia 63:4123–4133

Gilfillan SM, Sherk GW, Poreda RJ, Haszeldine RS (2017) Using noble gas fingerprints at the Kerr Farm to assess CO_2 leakage allegations linked to the Weyburn-Midale CO_2 monitoring and storage project. Int J Greenhouse Gas Control 63:215–225

Goertz-Allmann BP, Kühn D, Oye V, Bohloli B, Aker E (2014) Combining microseismic and geomechanical observations to interpret storage integrity at the In Salah CCS site. Geophys J Int 198(1):447–461

Golan M, Whitson CH (1991) Well performance, 2nd edn. Prentice Hall

Gollakota S, McDonald S (2014) Commercial-scale CCS project in Decatur, Illinois-Construction status and operational plans for demonstration. Energy Procedia 63:5986–5993

Hannis S, Chadwick A, Connelly D, Blackford J, Leighton T, Jones D, White J, White P, Wright I, Widdicomb S, Craig J (2017) Review of offshore CO_2 storage monitoring: operational and research experiences of meeting regulatory and technical requirements. Energy Procedia 114:5967–5980

Hansen H, Eiken O, Aasum TA (2005) Tracing the path of carbon dioxide from a gas-condensate reservoir, through an amine plant and back into a subsurface aquifer—case study: the Sleipner area, Norwegian North Sea. Society of Petroleum Engineers, SPE paper 96742. https://doi.org/10.2118/96742-ms

Hansen O, Gilding D, Nazarian B, Osdal B, Ringrose P, Kristoffersen JB, Eiken O, Hansen H (2013) Snøhvit: the history of injecting and storing 1 Mt CO_2 in the Fluvial Tubåen Fm. Energy Procedia 37:3565–3573

Hovorka SD, Meckel TA, Trevino RH, Lu J, Nicot JP, Choi JW, Freeman D, Cook P, Daley TM, Ajo-Franklin JB, Freifeild BM (2011) Monitoring a large volume CO_2 injection: year two results from SECARB project at Denbury's Cranfield, Mississippi, USA. Energy Procedia 4:3478–3485

Huang F, Bergmann P, Juhlin C, Ivandic M, Lüth S, Ivanova A, Kempka T, Henninges J, Sopher D, Zhang F (2018) The first post-injection seismic monitor survey at the Ketzin pilot CO_2 storage site: results from time-lapse analysis. Geophys Prospect 66(1):62–84

IEAGHG (2009) A review of the international state of the art in risk assessment guidelines and proposed terminology for use in CO_2 geological storage. IEA Greenhouse Gas R&D Programme, Report 2009-TR7

IEAGHG (2016) Offshore monitoring for CCS projects. IEA Greenhouse Gas R&D Programme, Report 2015/02

Ivanova A, Kashubin A, Juhojuntti N, Kummerow J, Henninges J, Juhlin C, Lüth S, Ivandic M (2012) Monitoring and volumetric estimation of injected CO_2 using 4D seismic, petrophysical data, core measurements and well logging: a case study at Ketzin, Germany. Geophys Prospect 60(5):957–973

Jenkins C, Chadwick A, Hovorka SD (2015) The state of the art in monitoring and verification—ten years on. Int J Greenhouse Gas Control 40:312–349

Jenkins C, Marshall S, Dance T, Ennis-King J, Glubokovskikh S, Gurevich B, La Force T, Paterson L, Pevzner R, Tenthorey E, Watson M (2017) Validating subsurface monitoring as an alternative option to surface M&V-The CO2CRC's Otway Stage 3 Injection. Energy Procedia 114:3374–3384

Johnsen K, Helle K, Røneid S, Holt H (2011) DNV recommended practice: design and operation of CO$_2$ pipelines. Energy Procedia 4:3032–3039

Johnson G, Raistrick M, Mayer B, Shevalier M, Taylor S, Nightingale M, Hutcheon I (2009) The use of stable isotope measurements for monitoring and verification of CO$_2$ storage. Energy Procedia 1(1):2315–2322

Jones DG, Beaubien SE, Blackford JC, Foekema EM, Lions J, De Vittor C, West JM, Widdicombe S, Hauton C, Queirós AM (2015) Developments since 2005 in understanding potential environmental impacts of CO$_2$ leakage from geological storage. Int J Greenhouse Gas Control 40:350–377

Kim S, Hosseini SA (2014) Above-zone pressure monitoring and geomechanical analyses for a field-scale CO$_2$ injection project in Cranfield, MS. Greenhouse Gases Sci Technol 4(1):81–98

Landrø M, Zumberge M (2017) Estimating saturation and density changes caused by CO$_2$ injection at Sleipner—Using time-lapse seismic amplitude-variation-with-offset and time-lapse gravity. Interpretation 5.2:T243–T257

Li H, Yan J (2009) Impacts of equations of state (EOS) and impurities on the volume calculation of CO$_2$ mixtures in the applications of CO$_2$ capture and storage (CCS) processes. Appl Energy 86(12):2760–2770

Liebscher A, Möller F, Bannach A, Köhler S, Wiebach J, Schmidt-Hattenberger C, Weiner M, Pretschner C, Ebert K, Zemke J (2013) Injection operation and operational pressure–temperature monitoring at the CO$_2$ storage pilot site Ketzin, Germany—Design, results, recommendations. Int J Greenhouse Gas Control 15:163–173

Lindeberg E (2011) Modelling pressure and temperature profile in a CO$_2$ injection well. Energy Procedia 4:3935–3941

Lumley DE (2001) Time-lapse seismic reservoir monitoring. Geophysics 66(1):50–53

Lüth S, Ivanova A, Kempka T (2015) Conformity assessment of monitoring and simulation of CO$_2$ storage: a case study from the Ketzin pilot site. Int J Greenhouse Gas Control 42:329–339

Maldal T, Tappel IM (2004) CO$_2$ underground storage for Snøhvit gas field development. Energy 29(9–10):1403–1411

Martens S, Kempka T, Liebscher A, Lüth S, Möller F, Myrttinen A, Norden B, Schmidt-Hattenberger C, Zimmer M, Kühn M (2012) Europe's longest-operating on-shore CO$_2$ storage site at Ketzin, Germany: a progress report after three years of injection. Environ Earth Sci 67(2):323–334

Martens S, Möller F, Streibel M, Liebscher A, Group TK (2014) Completion of five years of safe CO$_2$ injection and transition to the post-closure phase at the Ketzin pilot site. Energy Procedia 59:190–197

Mateeva A, Lopez J, Potters H, Mestayer J, Cox B, Kiyashchenko D, Wills P, Grandi S, Hornman K, Kuvshinov B, Berlang W (2014) Distributed acoustic sensing for reservoir monitoring with vertical seismic profiling. Geophys Prospect 62(4):679–692

Mathieson A, Midgley J, Dodds K, Wright I, Ringrose P, Saoul N (2010) CO$_2$ sequestration monitoring and verification technologies applied at Krechba. Algeria. The Leading Edge 29(2):216–222

Mayer B, Humez P, Becker V, Dalkhaa C, Rock L, Myrttinen A, Barth JAC (2015) Assessing the usefulness of the isotopic composition of CO$_2$ for leakage monitoring at CO$_2$ storage sites: a review. Int J Greenhouse Gas Control 37:46–60

McNab WW, Carroll SA (2011) Wellbore integrity at the Krechba carbon storage site, In Salah, Algeria: 2. Reactive transport modeling of geochemical interactions near the cement–formation interface. Energy Procedia 4:5195–5202 (GHGT-10)

Michael K, Golab A, Shulakova V, Ennis-King J, Allinson G, Sharma S, Aiken T (2010) Geological storage of CO$_2$ in saline aquifers—a review of the experience from existing storage operations. Int J Greenhouse Gas Control 4(4):659–667

Nazarian B, Held R, Høier L, Ringrose P (2013) Reservoir management of CO_2 injection: pressure control and capacity enhancement. Energy Procedia 37:4533–4543

Nazarian B, Thorsen R, Ringrose P (2018). Storing CO_2 in a reservoir under continuous pressure depletion—a simulation study. In: 14th greenhouse gas control technologies conference Melbourne 21–26 October 2018 (GHGT-14). Available at SSRN https://ssrn.com/abstract=3365822

Nordbotten JM, Celia MA, Bachu S (2005) Injection and storage of CO_2 in deep saline aquifers: analytical solution for CO_2 plume evolution during injection. Transp Porous Media 58(3):339–360

Oldenburg CM, Lewicki JL, Hepple RP (2003) Near-surface monitoring strategies for geologic carbon dioxide storage verification (No. LBNL-54089). Lawrence Berkeley National Laboratory (LBNL), Berkeley

Oldenburg CM, Bryant SL, Nicot J (2009) Certification Framework based on effective trapping for geological carbon sequestration. Int J Greenhouse Gas Control 3(4):444–457

Pawar RJ, Bromhal GS, Carey JW, Foxall W, Korre A, Ringrose PS, Tucker O, Watson MN, White JA (2015) Recent advances in risk assessment and risk management of geologic CO_2 storage. Int J Greenhouse Gas Control 40:292–311

Peaceman DW (2000) Fundamentals of numerical reservoir simulation, vol 6. Elsevier

Pruess K, García J, Kovscek T, Oldenburg C, Rutqvist J, Steefel C, Xu T (2004) Code intercomparison builds confidence in numerical simulation models for geologic disposal of CO_2. Energy 29(9–10):1431–1444

Ringrose PS, Meckel TA (2019) Maturing global CO_2 storage resources on offshore continental margins to achieve 2DS emissions reductions. Scientific Reports 9:17944. https://doi.org/10.1038/s41598-019-54363-z

Ringrose PS, Mathieson AS, Wright IW, Selama F, Hansen O, Bissell R, Saoula N, Midgley J (2013) The In Salah CO_2 storage project: lessons learned and knowledge transfer. Energy Procedia 37:6226–6236

Ringrose P, Greenberg S, Whittaker S, Nazarian B, Oye V (2017) Building confidence in CO_2 storage using reference datasets from demonstration projects. Energy Procedia 114:3547–3557

Romanak KD, Bennett PC, Yang C, Hovorka SD (2012) Process-based approach to CO_2 leakage detection by vadose zone gas monitoring at geologic CO_2 storage sites. Geophys Res Lett 39(15)

Rutqvist J (2012) The geomechanics of CO_2 storage in deep sedimentary formations. Geotech Geol Eng 30(3):525–551

Sharma S, Cook P, Jenkins C, Steeper T, Lees M, Ranasinghe N (2011) The CO2CRC Otway project: leveraging experience and exploiting new opportunities at Australia's first CCS project site. Energy Procedia 4:5447–5454

Singh VP, Cavanagh A, Hansen H, Nazarian B, Iding M, Ringrose PS (2010) Reservoir modeling of CO_2 plume behavior calibrated against monitoring data from Sleipner, Norway. In SPE annual technical conference and exhibition. Society of Petroleum Engineers. https://doi.org/10.2118/134891-MS

Span R, Wagner W (1996) A new equation of state for carbon dioxide covering the fluid region from the triple point temperature to 1100 K at pressures up to 800 MPa. J Phys Chem Ref Data 25:1509–1596

Span R, Gernert J, Jäger A (2013) Accurate thermodynamic-property models for CO_2-rich mixtures. Energy Procedia 37:2914–2922

Thibeau S, Gapillou C, Marblé A, Urbanczyk C, Garnier A (2017) The Rousse CO_2 storage demonstration pilot: hydrogeological impacts of hypothetical micro-annuli around the cements of the injector well. Pet Geosci 23(3):363–368

Van der Tuuk Opedal N, Torsæter M, Vrålstad T, Cerasi P (2014) Potential leakage paths along cement-formation interfaces in wellbores; Implications for CO_2 storage. Energy Procedia 51:56–64

Vasco DW, Rucci A, Ferretti A, Novali F, Bissell RC, Ringrose PS, Mathieson AS, Wright IW (2010) Satellite-based measurements of surface deformation reveal fluid flow associated with the geological storage of carbon dioxide. Geophys Res Lett 37(3)

Ringrose P, Furre AK, Bakke R, Dehghan Niri R, Paasch B, Mispel J, Bussat S, Vinge, T, Vold L, Hermansen A (2018) Developing optimised and cost-effective solutions for monitoring CO$_2$ injection from subsea wells. In: 14th greenhouse gas control technologies conference Melbourne, pp 21–26

White D, Harris K, Roach L, Roberts B, Worth K, Stork A, Nixon C, Schmitt D, Daley T, Samson C (2017) Monitoring results after 36 ktonnes of deep CO$_2$ injection at the Aquistore CO$_2$ storage site, Saskatchewan, Canada. Energy Procedia 114:4056–4061

Will R, El-Kaseeh G, Jaques P, Carney M, Greenberg S, Finley R (2016) Microseismic data acquisition, processing, and event characterization at the Illinois Basin–Decatur Project. Int J Greenhouse Gas Control 54:404–420

Williams GA, Chadwick RA (2017) An improved history-match for layer spreading within the Sleipner plume including thermal propagation effects. Energy Procedia 114:2856–2870

Williams GA, Chadwick RA, Vosper H (2018) Some thoughts on Darcy-type flow simulation for modelling underground CO$_2$ storage, based on the Sleipner CO$_2$ storage operation. Int J Greenhouse Gas Control 68:164–175

Worth K, White D, Chalaturnyk R, Sorensen J, Hawkes C, Rostron B, Johnson J, Young A (2014) Aquistore project measurement, monitoring, and verification: from concept to CO$_2$ injection. Energy Procedia 63:3202–3208

Wright IW, Ringrose PS, Mathieson AS, Eiken O (2009) An overview of active large-scale CO$_2$ storage projects. In SPE international conference on CO$_2$ capture, storage, and utilization. Society of Petroleum Engineers

Zhang M, Bachu S (2011) Review of integrity of existing wells in relation to CO$_2$ geological storage: what do we know? Int J Greenhouse Gas Control 5(4):826–840

Zweigel P, Arts R, Lothe AE, Lindeberg EB (2004) Reservoir geology of the Utsira Formation at the first industrial-scale underground CO$_2$ storage site (Sleipner area, North Sea). Geol Soc Lond Spec Pub 233(1):165–180

Chapter 4
Future of CCS—What Happens Next?

In this short introduction to CO_2 storage technology, we have demonstrated that engineered geological storage of CO_2 is an active and established technology, using various examples from early-mover projects. We have focussed on the case of saline aquifer injection, using examples from the long-running offshore projects at Sleipner and Snøhvit in Norway and the onshore industrial-scale projects at In Salah in Algeria, Quest in Canada and Decatur in the USA. Useful insights from smaller pilot projects were also drawn, including the Ketzin CCS project in Germany (which has now moved into the post-closure phase) and the CO_2 injection test sites at Otway (Australia), Cranfield (USA) and Lacq-Rousse (France). This was not intended as a comprehensive review, but rather a selection of insights gained as a basis for understanding the concepts and methods involved. Globally we currently have 19 large-scale CCS facilities in operation with an installed capture capacity of 36 Mtpa (GCCSI 2019), so the technology can be considered as in operation. Note that most of these projects use oilfields as the storage domain, via the process of enhanced oil recovery (EOR) using CO_2 as the injection fluid. Storage as part of CO_2 EOR projects has not been covered in this review but is extensively reviewed elsewhere (e.g. Hitchon 2012; Eide et al. 2019).

In the IEA (2016) review of CCS, called '20 Years of Carbon Capture and Storage' (marking the 20th anniversary of the Sleipner CCS project), highlighted both the successes of and challenges for CCS as a climate mitigation technology, noting that:

- Two decades of CCS experience has led to a growing recognition by climate experts of the value and potential of the technology;
- However, this increased recognition of CCS has not been matched with increased support. CCS deployment has been hampered by fluctuating policy frameworks and lack of financial support.
- Despite this, most climate-change mitigation models agree that CCS is central to achieving <2 °C global warming pathway. CCS is an essential part of the least-cost portfolio for the power sector and the main mitigation solution in many industrial sectors.

P. Ringrose, *How to Store CO2 Underground: Insights from early-mover CCS Projects*, SpringerBriefs in Earth Sciences, https://doi.org/10.1007/978-3-030-33113-9_4

So, CCS is a proven and tested technology which is essential to meeting the greenhouse gas reduction goals implied by the Paris agreement, and yet it is progressing much slower than needed. CCS is anticipated to support approximately 13% of total cumulative emissions reductions by 2050, which implies around 120 Giga tonnes (Gt) of cumulative CO_2 reduction using CCS by 2050 (IEA 2015) and an increase in global capture rate (currently at around 40 Mtpa) by a factor of 150 by 2050.

Is this possible? Technically, yes—there is ample geological capacity for required storage projects (Ringrose and Meckel 2019). The main problem is a socio-economic one—the costs of CCS are perceived to be too high and the benefits are perceived to be too low. How then might this perception change? Below, we identify some of the main arguments and drivers for accelerating CCS technology in the near future.

The climate-mitigation value proposition: Within the total portfolio of climate-mitigation actions, CCS lowers the costs of achieving the required emissions reductions in the <2 °C warming scenario (compared to non-CCS pathways), and many would argue the goal is not achievable without CCS. The societal value of CO_2 storage in deep geological formations also needs emphasizing. Put simply, CO_2 storage is a lot safer and better than putting the same CO_2 into the atmosphere. Furthermore, a single CO_2 injection well disposing of around 1 Mt of CO_2 per year represents a very significant emissions reduction (the single injection well at Sleipner well corresponds to approximately 10% of all Norwegian road traffic emissions). So, as an emissions-reduction activity geological CO_2 storage is highly effective.

The carbon price effect: As the cost of emitting to atmosphere increases, CCS will become increasingly attractive. The cost of emitting CO_2 around the world is highly variable, varying from zero in many places to a growing number of nations with emissions prices in the range of 20–60 US\$/tCO_2e (carbonpricingdashboard.worldbank.org). As the carbon price rises, CCS projects become more likely to be implemented. Furthermore, the cheaper capture technologies (namely CO_2 capture from gas processing, bioethanol production and fertilizer plants) are likely to proceed first, becoming economic with a carbon price of around 50 US\$/tCO_2e (GCCSI 2017). CCS projects in the fossil-fuel power sectors and in the iron, steel and cement industrial sectors are only likely to proceed with carbon prices of over 100 US\$/tCO_2e.

The infrastructure effect: CCS projects represent large infrastructure projects, and the initial capital investments (typically several 100's MUSD) only start to make sense once the 'economy of scale' begins to kick in. In fact, the Sleipner CO_2 capture and processing facility (platform-based), has already started to act as a CCS hub by processing gas streams from neighbouring fields (Ringrose 2018). Establishing common CO_2 transport networks (e.g. Stewart et al. 2014) and developing novel ways to use existing oilfield infrastructure to enable cost effective CCS (e.g. Scafidi and Gilfillan 2019) are likely to be key to stimulating significant scale-up of CCS. Currently, several parallel initiatives in the countries bordering the North Sea basin (Norway, UK, Netherland, Denmark and Germany) suggest that the North Sea could become the world's first integrated CCS hub.

The negative emissions driver: As the urgency of getting the human population to change its behavior of continuing greenhouse gas emissions to the atmosphere grows, the need to adopt 'negative emissions technology' becomes more pressing. Sucking CO_2 out of the atmosphere is one way of compensating for excess emissions in the past (IPCC 2018), and the two main technologies involved are direct air capture (DAC) and CO_2 capture from the combustion of biomass feedstock (BECCS). Both technologies only make sense if the captured CO_2 is permanently isolated from the atmosphere, and so geological storage of CO_2 will be essential to the deployment of negative emissions technology.

If some, or all, of the above factors begin to gain traction, then we might indeed see the significant acceleration in CCS project deployment that is required to meet the global climate mitigation objectives. Whatever happens, one thing is clear: Engineered geological storage of CO_2 is an available technology and a highly beneficial one. Putting CO_2 in deep geological formations is a lot safer than emitting the same CO_2 into the atmosphere. CO_2 storage is a technology that is ready to be used—the main remaining question is 'does our society want to use it?' I hope so, and I hope this introduction to the topic will inspire you to engage with this technology.

References

Eide LI, Batum M, Dixon T, Elamin Z, Graue A, Hagen S, Hovorka S, Nazarian B, Nøkleby PH, Olsen GI, Ringrose P, Vieira RAM (2019) Enabling large-scale Carbon Capture, Utilisation, and Storage (CCUS) using offshore carbon dioxide (CO_2) infrastructure developments—a review. Energies 12(10):1945

GCCSI (2017) Global costs of Carbon capture and storage—2017 update. Global CCS Institute

GCCSI (2019) GCCSI CO2RE database: 2019. Global CCS Institute, 2019. https://co2re.co

Hitchon B (ed) (2012) Best practices for validating CO_2 geological storage: observations and guidance from the IEAGHG Weyburn-Midale CO_2 monitoring and storage project. Geoscience Publishing

IEA (2015) Carbon capture and storage: the solution for deep emissions reductions. International Energy Agency Publications, Paris

IEA (2016) 20 years of carbon capture and storage: accelerating future deployment. https://www.iea.org/publications

IPCC (2018) Summary for policymakers. In: Masson-Delmotte V, Zhai P, Pörtner H-O, Roberts D, Skea J, Shukla PR, Pirani A, Moufouma-Okia W, Péan C, Pidcock R, Connors S, Matthews JBR, Chen Y, Zhou X, Gomis MI, Lonnoy E, Maycock T, Tignor M, Waterfield T (eds) Global warming of 1.5°C. An IPCC special report on the impacts of global warming of 1.5°C above pre-industrial levels and related global greenhouse gas emission pathways, in the context of strengthening the global response to the threat of climate change, sustainable development, and efforts to eradicate poverty. https://www.ipcc.ch/sr15/

Ringrose PS (2018) The CCS hub in Norway: some insights from 22 years of saline aquifer storage. Energy Procedia 146:166–172

Ringrose PS, Meckel TA (2019) Maturing global CO_2 storage resources on offshore continental margins to achieve 2DS emissions reductions. Scientific Reports 9:17944. https://doi.org/10.1038/s41598-019-54363-z

Scafidi J, Gilfillan SM (2019) Offsetting Carbon Capture and Storage costs with methane and geothermal energy production through reuse of a depleted hydrocarbon field coupled with a saline aquifer. Int J Greenhouse Gas Control 90:102788

Stewart RJ, Scott V, Haszeldine RS, Ainger D, Argent S (2014) The feasibility of a European-wide integrated CO_2 transport network. Greenhouse Gases Sci Technol 4(4):481–494